Design Research Foundations

The goal of the series is to provide a platform for publishing state of the art research on foundational issues in design and its applications in industry and society. Suitable topics range from methodological issues in design research to philosophical reflections on the specificities of design rather than actual design work or empirical cases only. The definition of design behind the series is inclusive. In terms of disciplines, it ranges from engineering to architecture. In terms of design work, it ranges from conceptual issues in design through design experiments and prototypes to evaluative studies of design and its foundations.

Proposals should include:

- A proposal form, as can be found on this page
- A short synopsis of the work or the introduction chapter
- The proposed Table of Contents
- The CV of the lead author(s)
- If available: one sample chapter

We aim to make a first decision within 1 month of submission. In case of a positive first decision the work will be provisionally contracted: the final decision about publication will depend upon the result of the anonymous peer review of the complete manuscript. The series editors aim to have the complete work peer-reviewed within 3 months of submission.

The series discourages the submission of manuscripts that contain reprints of previous published material and/or manuscripts that are below 150 pages/75,000 words.

For inquiries and submission of proposals authors can contact the series editors, Pieter Vermaas via: p.e.vermaas@tudelft.nl; or Ilpo Koskinen via: ilpo.koskinen@polyu.edu.hk

More information about this series at http://www.springer.com/series/13775

Annie Gentes

The In-Discipline of Design

Bridging the Gap Between Humanities and Engineering

Springer

Annie Gentes
Codesign Lab, I3
Telecom ParisTech
Paris, France

ISSN 2366-4622 ISSN 2366-4630 (electronic)
Design Research Foundations
ISBN 978-3-319-65983-1 ISBN 978-3-319-65984-8 (eBook)
DOI 10.1007/978-3-319-65984-8

Library of Congress Control Number: 2017951023

Printed on acid-free paper

This Springer imprint is published by Springer Nature
The registered company is Springer International Publishing AG
The registered company address is: Gewerbestrasse 11, 6330 Cham, Switzerland

Acknowledgments

A Place to Start

This book started in a garage in Silicon Valley thanks to Alex Pang and his family who hosted me for a month when I first arrived for a sabbatical at Stanford. Alex is an amazing writer and thinker of our contemporary digital lives as well as a renowned futurologist. Starting this book in a garage was of course very emblematic of the local culture. More importantly, this sabbatical year as Alex aptly describes it in his book *Rest* was a turning point in my life. At long last, I could start a long reflexive work, stepping back from the everyday publish and perish (and teach, manage, and organize). Thank you so much Alex and Heather.

Mentors

I am extremely grateful for Gerard Pogorel's advice to look into the conception and design of services at the beginning of my career. He has been a true mentor and helped me, through long discussions, to question and explore design practices and productions. Many people from Telecom ParisTech also included me in their projects. I want to thank Jean-Claude Moissinac who was the first to trust that we could bring together engineering, design, social, and human sciences. He opened the way to many other wonderful collaborations with Isabelle Demeure, Catherine Pelachaud, Michel Simatic, and Eric Gressier-Soudan (CNAM) who have been regularly quoted in this book, as well as designers from ENSCI, ENSAD, and Strate School of Design and in particular Frederique Pain who is a wonderful partner in research.

Languages

French is my native tongue, while English is my adopted tongue. Translation is at the core of my experience. However, I despair of being ever really good at both. Kate Meyer's English language editing was outstanding in improving the texts. She read every sentence I wrote, fixed the ones she could, and was tactful about the ones beyond redemption.

Frontiers

I want to give special credit to the artists and engineers, designers, social scientists, and humanities scholars, who have been mentioned many times in this book. They all brought me into their worlds where I could wonder at the richness of their contributions to knowledge and aesthetics. During my sabbatical year, I met some great people at Stanford who showed me a wide variety of design research which helped me enormously in deciding where I personally stood: in particular, Tim McAloon who does research in service and sustainable design and Mark Nelson and Margarita Quihuis who not only believe in peace innovation but also supported me when I faced major bureaucratic crisis. You have been incredible friends and a great inspiration.

Teams

Writing is a lonely process where we face ourselves and our failings. However, being part of a team changes everything. This book would not have been possible without the support of members of the CodesignLab and of the MINES ParisTech Chair on Theories and Methods of Conception (TMC). I especially want to thank Armand Hatchuel for his wise recommendations, his encouragements, and the space of intellectual freedom that he has opened with his collaborators. It is thanks to him that I met so many good people who have helped me again and again like Anne-Françoise Schmid, Mathias Béjean, Paris Chrisos, and Muriel Mambrini. My heartfelt thanks also go to the pioneers of the CodesignLab including Aude Guyot and Camille Jutant, Tiphaine Kazi-Tani, Cédric Mivielle, and Max Mollon and all the members past and present of the CodesignLab who have devoted long hours to discussing ideas and eating chocolate.

Readers

We probably all need a first enthusiastic reader to muster the courage to jump onto the first blank page and, more importantly, to go on writing. My heartfelt thanks go to Ted Selker, an IBM fellow and a fantastic pluridisciplinary scientist, whom I met at Carnegie Mellon University at Silicon Valley. He spent many hours sharing with me his amazing stories about all the great designs that he invented and pushed me to question every aspect of my own experience. His support was invaluable at a time of hesitation and doubt. Warren Sack was also among the first readers whose scope of scientific knowledge as well as attention to details was an invaluable contribution. I am also especially grateful for Ilpo Koskinen and Pieter Vermaas' patience and enthusiastic support for the book. They, with the anonymous scientific reviewers, provided invaluable advice on making the book more focused and, I hope, accessible to readers.

Family

I am forever indebted to Benjamin and Cecilia on so many levels that it is hard to know where to start. I would have done nothing without their love and intelligence. I have tried a lot of my ideas, questions, and prototypes with them. They have been more than patient about my obsessive interest in design. Above all, they have given so much affection without which I would have found difficult to design my own existence. This book is dedicated to them.

Contents

Chapter 1
Design as Meaning and Form Making: An Introduction

1.1 Book General Viewpoint and Goal

What does it take to produce an original work of science, art, or design? Are inventors and designers all "Leonardos" facing the daunting task of connecting all bodies of knowledge? How are the disciplines of conception related to social sciences and the humanities? This book tries to tackle these questions through the analysis of art and design projects as well as technological research programs that not only devise new technologies but also explore their potential in unexpected ways.

It all started because I was a social scientist within an electrical engineering and computer science school. I was intrigued by what my colleagues did and thus started looking into their activities, but very quickly I ended up helping with some of their projects. It was fascinating and exhilarating to be part of creative teams. However, being no engineer myself, I had to find what concepts or methods from my original disciplines could really help. How could I contribute to the invention of information technologies with a background in English and American literatures I wrote a memoir on Tolkien's Lord of the Rings and a PhD on public communication? At first (and probably second) sight, there was nothing in common between, on the one hand, my education in the humanities and in information and communication sciences, and, on the other hand, engineering research. Still, at "third sight", I found that some of the things that I had learned along the way could actually help design artifacts and in particular new media[1] or, as I call them "reflective technologies". It led me to explore a certain perspective on design (the specific practices) and design (the process of conception) that was based on this double academic education. Before going into the concepts and methods along with the examples described in the different chapters in more details, I want to give a quick overview of how projects and analyses are generally presented in the book.

[1] Manovich (2002).

A. Gentes, *The In-Discipline of Design*, Design Research Foundations,
DOI 10.1007/978-3-319-65984-8_1

1.1.1 Examples and Context

The choice of examples ranging from research to design and art is done on purpose. Even though I want to do justice to the diversity of activities, I also want to uncover the common points between different creative activities whatever their object and goal. Indeed, I think we need to develop an understanding of design/conception in its different forms.

Research examples are mostly picked up from collaborations of the Codesign lab[2] of Telecom ParisTech[3] with public or industrial tech labs over a span of nearly 20 years. The Codesign lab includes social scientists, humanities scholars, and designers, who work with teams in computer science and signal and image processing on questions such as distributed architecture, mobile and pervasive computing, or intelligent virtual agents. The sample of projects is, I think, representative of what is at stake in the field of information and communication technologies. Moreover, these collaborations gave me a chance to observe and test design methods over long periods of time. First, the observation/participation could take place at the very beginning of the projects and until their completion, which is a condition for the success of a longitudinal approach. Second, the groups that participated in the projects accepted that I could carry out an in-depth empirical study, which was another necessary condition of sustaining a comprehensive approach to design practices at work. Third, the team members (whether engineers or artists) were already engaged in a reflection on the use of different tools in their practices and welcomed a reflection on the various ways that engineering, and social and human sciences could work together.

The art examples come from various sources and were often suggested by artists that I met or worked with (Carol-Ann Braun, Ludovic de Vita, Agnes de Cayeux, Antoine Schmid, amongst others). At the beginning of my career, I was interested in and got the opportunity to study and contribute to some new media art projects as well as to teach engineering students how to develop their own artistic and creative skills. The examples are far from representing the full gamut of new media art but they bring to the foreground properties of the digital fabric and also of the internet as a media that supports unprecedented aesthetic experiences.

Design examples are mostly related to the engineering research examples in which designers played an important role through the Codesign lab. They also come from PhD students in design (Aude Guyot, Cedric Mivielle, Max Mollon, in particular) and from designers from various design institutions in France, in particular

[2]The Codesign lab of Telecom Paristech is a pluridisciplinary lab specialized in the analysis of design and its diversity of practices. I created it in 2000 with a collaboration of ENSCI (Ecole Nationale Supérieure de Création Industrielle) and the department of computer science of Telecom Paristech. It was developed with the help of Armand Hatchuel of Mines Paristech, another engineering school which has developed a groundbreaking research in design theory through the industrial chair "Theory and methods of design".

[3]Telecom Paristech is an engineering school in Paris founded in 1878 and specialized in communication and information technologies. http://www.telecom-paristech.fr/

ENSCI (the Ecole Nationale Supérieure de Création Industrielle), ENSAD (Ecole Nationale Supérieure des Arts Décoratifs) Strat College of Design but also Valence Art and Design School. I also met designers from the Royal College of Art and Goldsmith College in London, as well as designers working at or with Microsoft Research Socio-Digital System Group (headed by Abigail Sellen and Richard Harper) or with Sony Music Lab in Paris. I chose these design examples because they presented a variety of starting points: either an emerging technology that was forging its identity, or a critical stand that forced spectators to suspend, for a while, their scientific enthusiasm, or an exploration of alternative uses of digital media.

1.1.2 People

As the reader will see, this book is the result of hundreds of interviews with creators from very different backgrounds. Here, I want to mention people that I do not refer to in this book but who actually showed me a great deal about invention. In my PhD that dealt with public communication in Departments of State, I had the opportunity to meet the professionals who mastered the delicate art of creating a strategy of communication where none was supposed to be, who designed not only the strategy but also the dramaturgy of the State, and co-created discourses with the different actors of the political arena.[4] At the time, departments of State were considered and criticized as bureaucracies (they still are) but I was more interested in how they managed to be innovative. Today, a lot of the literature relates to social innovation and design. But 30 years ago, I witnessed, first hand, what it meant to create new social and political paths to solve problems. Of course, working with engineering researchers, designers, and artists, has made it easier to look at radical innovation. But these very constricted situations of creation also gave me a chance to understand what I like to call "extreme design", that is design that takes place in organizations or times that are not specially geared for innovation. This book is therefore dedicated to creators, whether they are designers, engineers, researchers in engineering, social scientists, or more generally public and private actors, because they design new artifacts and situations, concepts and methods in sometimes complicated circumstances. I have written this book especially for people who are concerned not only with the technical, but social, political, and aesthetical aspects of their invention.

[4] Gentes (1996).

1.1.3 Hybrid Methods

To study the design of new artifacts whether artistic or other, I hijacked the interdisciplinarity which is at the core of information and communication sciences.[5] This discipline uses the humanities as well as social sciences to understand composite objects[6] made of different semiotic materials and being performed in a variety of situations. In other words, the reader will find that to do my research, I studied

- the discourses on design (design objects, design processes) or of design (actors' interactions), either in texts or in situations of design,
- the productions of design either linguistic, iconic, tangible, and their role in design/conception.

For example, inspired by the humanities, I looked at the role of corpuses in design. I could then observe that both humanities scholars and designers rely on historical perspective, media studies, and semiotic methodologies to establish connections, define, and contrast families of objects. I could also see how both human scientists and designers are concerned with language, communication, reception, and can use translation techniques to move from one set of media to another. Combining both language and visual analyses I also noticed that both were keen on intermediality, that is to say meaning making built from different semiotic categories, supported by different media.

I also borrowed from ethnographic and sociological methods to better understand the circulation of information and the role of actors and structures. I used participative observation, in-depth interviews, and I followed documents production, distribution, and consumption. I also related the actors' activities and productions to their institutional and political framework: for example, research policies, museum legitimacy, or art worlds. Again, my goal was not only to use the interdisciplinarity of information and communication sciences as a tool to address the design situation. I also wanted to check if and how the same interdisciplinary background was part of the discipline of design itself. I could indeed see how designers base their work on an analysis of activities, a better understanding of their users, a careful understanding of complex social situations.[7,8]

To sum up, this book tries to achieve a pragmatic approach to the discipline of design by using the humanities and social sciences and focusing both on the designed objects and the analysis of creators' activities. In the process of writing this book, each discipline has been used not only to complement each other, for example bringing the analysis of styles to better understand the use of inventions, but also to under-determine each other, that is to say to question the limits of their understanding of design. Another way to say it, is that I have tried to bring together

[5] Davallon (2004).

[6] Souchier et al. (2003).

[7] Akrich (1990).

[8] Akrich (1992).

design studies and design science:on the one hand the analysis of cultural and historical artifacts, and on the other the analysis of rationales of generative practices.

1.1.4 Structure

I have tried to give each chapter a similar structure based on the presentation of detailed examples of research, design, or art, then elaborating on more theoretical reflections. In almost every chapter, two types of conclusions are derived from the examples:

- a first level considers the lessons that we can derive from these examples in terms of design practice. The first level is therefore deliberately pragmatic. While this book does not encompass all design practices, the examples are somehow emblematic of the kind of challenges that designers are faced with. Therefore, the different methods are not necessarily to be taken as tricks of the trade but rather as dealing with design issues such as naming, or borrowing, or dealing with function and/or contents.
- the second level starts a theoretical discussion. This theoretical discussion is based on a semiotic and media studies point of view of design. I contend that design not only creates new meaning but builds a specific plane of composition that is both related to and independent from the experienced world. We must understand this meaning-making process with an emphasis on the "making" and a focus on its material and conceptual tools. This might complement the reader's knowledge of other fields that deal with design theory such as design management or human computer interaction. My goal is to explore the complexity of design/practice and of design/conception by reporting what the methods of humanities help us uncover.

1.2 Definitions of Design: The Challenge of New Beginnings

While this book belongs to the field of research known as the sociology of science[9] as it relies heavily on its findings and adds to the understanding of how science is made, it is primarily a book on design conceived as a series of activities that produce new meaning through the realization of new tangible artifacts, services, situations.[10]

Because in this book design is considered as a meaning and form making activity, I looked at situations or artifacts when a definition is yet to be given, or is questioned, when the identity is unknown or challenged, and assumptions are wide open

[9] Latour (1987).

[10] Buchanan (2001).

or contested. The book is therefore a book about beginnings: how people start thinking about something that they cannot yet name, how they try to foretell the future of their work, how they plan the first steps of their invention, how they involve different stakeholders in a debate.

As usual in books dedicated to design, a definition needs to be made of how the word design is to be understood. Historically speaking, design inherits both from art and industry. From art, design knows that it produces an aesthetic experience in the beholder or user of its artifacts. This part of design is shared with artists. The main model of design in this context is that of the "matrix" that gradually shapes the invention. The "matrix" metaphor emphasizes the spatial and holistic plane of creation. From industry, design knows that it produces objects with a function in as much as objects organize the relationship of man to the world. This part of design is shared with engineers. The model of design in this context is also specific and relies on the "project", a temporal division of work. I think that creators, whatever their field, are always caught between these two conceptions of the design activity, and often tend to polarize between the two poles: holistic reasoning/sequential reasoning in relation to their specific training and goal. Because of my background, and because I found that extensive research had been done on the concept of "project of design",[11] I have focused more on the holistic dimension of design.

In any event, design in this book qualifies both specialized practices of designers and conception in general. First, design as creative practices that are specific to designers is sometimes alluded to by using the "design/practice" abbreviation. In this instance, it means theories or methods are circumscribed by what professional designers and theoreticians consider as the specificity of their discipline. It helps me to contrast engineers and designers' ways of doing things. Second, design as conception, whether or not it is pursued by designers, is sometimes introduced through the expression "design/conception". Here, the focus is on how very diverse creative activities all strive to invent a new artifact, language, service, or to express it in more general terms: produce an X. Robust models of design have been developed from a logical point of view (for example CK theory[12]), a psychological and cognitive point of view (for example Norman's approach[13]), and an organizational point of view (for example the design thinking school[14]). My proposition is to envision, through all the examples coming from different fields, design as a meaning and form making, media related, activity, that draws not only from the logical and material construct, but from the aesthetic, the symbolic, and the communicational construct.

[11] Findeli and Coste (2007).

[12] Hatchuel et al. (2012).

[13] Norman (2002).

[14] Cross (2011).

1.3 Epistemology of Design: Building the Future

If science is not only about analyzing what is but actually building its object, if it is about "this orientation towards the future – towards 'what might be'", then we need to think about ways to understand the place of design in research and invention. Steven Dow, Wendy Ju and Wendy MacCay[15] studied the different ways design is used in engineering research, but also stressed that a whole new perspective needs to be put forward that associates science and design/practice in very novel ways. The first consequence is that we need methods, concepts, and tools that deal with things that are not yet there. Within this perspective, design and science are no longer considered as one following the other, that is as research bringing new material which is then exploited and deployed by design. Research and design are considered together as expansive and symbolic practices before being considered as specific practices. Design/practice is therefore considered emblematic of design/conception.

1.3.1 Abduction and Formal Practice

In this respect, the concept of abduction introduced by Peirce comes to the forefront of my reflection as a way to better understand science and design/conception. I will develop this idea later in the book but I want to stress now that abduction is, from a Peircian viewpoint, the only really creative part of science. This leads me to study any practice and production that is generative of new concepts, artifacts, and knowledge. But abduction in this context covers not only concepts but the series of practices and productions that assemble elements in unexpected ways to create new things. This trend of research has also been explored by other authors in relation to design like Shank,[16] Jonas and Chow,[17] or Jutant, Gentes, Béjean.[18] Here I want to look at the formal and situated means of abduction in engineering and design. I think it is important to analyze how formats, structures, languages, contribute to identifying and designing a radical unknown. The abductive practices of inventors and creators need to be understood from a formal viewpoint. Throughout the book, I have hunted for the manifestations of these formal practices that bring together materials in unexpected ways. While this research elaborates on a whole field of reflections based on Simondon,[19] Simon,[20] Leroi-Gourhan,[21] I have tried to look at

[15] Dow et al. (2013).

[16] Shank (2001).

[17] Chow and Jonas (2010).

[18] Jutant et al. (2013).

[19] Simondon eand Hart (2001).

[20] Simon (1996).

[21] Leroi-Gourhan and White (1993).

the invention and evolution of technologies as poetic practices. Indeed, I think that to invent a new thing, one must identify an autonomous plan of conception that is only indirectly related to a teleological perspective of action on the world. This autonomous poetical plan of conception can be a moodboard, a narrative, a scenario, etc. In any event, it has rules of its own that rely on the generative and coherent properties of the media. Creators not only design a radical unknown they also create this poetical space where a tangible unknown can be explored.

Finally, we need to acknowledge designers' claim that they are inter-disciplinary because they engage with technology, culture, society and science. Abduction could then be understood as the epistemological basis for what I suggest we call an "in-discipline" of design, a concept that I develop in the last chapter of this book.

1.3.2 Introducing Humanities to Design: Following Foucault and Peirce

Social sciences have been used in engineering sciences to elaborate more complex artifacts. They basically bring the activities of the users into the technological model. But I believe that the humanities can also bring their perspective which is based on the discovery and analysis of the new, the original, and the specific, Stolterman speaks of the "ultimate particular" which is the episteme of design.[22] The ultimate particular "is a singular and unique composition or assembly". For Stolterman, "design is a process of moving from the universal, general, and particular (protocols-rules of relationships and prescriptive specifications) to the ultimate particular – the specific design". Humanities scholars have developed ways to face the challenge of understanding what has no previous equivalent. Even in the Age of Mechanical Reproduction,[23] visual art and literature produce not only new versions, but new genres, new paradigms that need to be discovered and analyzed. The methods and concepts of the humanities were developed to deal with these new forms, and explore in what way they may help us build the future. Buchanan leads us in that direction when, in an article about the history of learning, he reminds us that design after centuries of functionalism could well go back to the humanities following the lost program of Bacon to turn our knowledge towards "artificial things".[24] A way to look at this invitation is to look at what types of knowledge are concerned. What is striking today in design schools is the combination of social sciences and their focus on the user, as well as practical knowledge about techniques and materials, but also an emphasis on critical and historical background in fine arts and cultural productions in general. In this book, I want to provide the reader with a sense of what it means to borrow from all these academic traditions and in particular the humanities.

[22] Nelson and Stolterman (2012).

[23] Benjamin (2010).

[24] Buchanan (2001).

To do so, I want to follow Foucault's definition of human sciences in *The order of Things: An Archeology of Human Sciences*, 1966.

Foucault defines "three epistemological regions" that study what being human means.

> The domain of the human sciences is covered by three 'sciences' or rather by three epistemological regions, all subdivided within themselves, and all interlocking with one another; these regions are defined by the triple relation of the human sciences in general to biology, economics, and philology.[25]

Foucault delineates a first group of scientific perspectives around psychology and the embodied functions of humans. "Thus one could admit that the 'psychological region' has found its locus in that place where the living being, in the extension of its functions, in its neuro-motor blueprints, its physiological regulations, but also in the suspense that interrupts and limits them, opens itself to the possibility of representation". Centered on the individual, this field of research can be applied to design in as much as conception can be studied from a creativity viewpoint. Questions are raised about the creators' psychology, their cognitive patterns, their sensibility and how they address users' passions, either positive or negative.

Second, Foucault describes how individuals are always part of groups who structure the way they live in society: "In the same way, the 'sociological region' would be situated where the laboring, producing, and consuming individual offers himself a representation of the society in which this activity occurs, of the groups and individuals among which it is divided, of the imperatives, sanctions, rites, festivities, and beliefs by which it is upheld or regulated." Social sciences are focused on situated activities. The relevance of social sciences for design is to consider design as a social activity. Indeed, much of the research on design observes how an organization or a group can support creative skills and activities.

Lastly, Foucault argues that, "in that region where the laws and forms of a language hold sway, but where, nevertheless, they remain on the edge of themselves, enabling man to introduce into them the play of his representations, in that region arise the study of literature and myths, the analysis of all oral expressions and written documents, in short, the analysis of the verbal traces that a culture or an individual may leave behind them". This region is that of humanities as a whole field of enquiry, different from social sciences in that it focuses not on the activity but on the productions and the representations of the activity. Design studies represents this branch in so far as they study the productions of design over time and space, but also because some of the research has come up with a communicational model of design. Elaborating on these researches,[26] it is the hypothesis of this book that the methods commonly used to study the "traces" of culture are also methods assessing the originality and innovativeness of human productions (and not the repetitiveness of human practices) as well as design methods.

---

[25] Foucault (1994).

[26] In particular Crilly, Chow, Jonas.

In this Foucaldian paradigm, design belongs to the three regions of the human sciences not only because it can be studied from these perspectives but also because we can reformulate his proposition in semiotic terms that focus on the generative and expansive dimensions of human activities. Hence, the central place of Peirce in my work.

Peirce produced a philosophical model of meaning based on a series of triadic complementary properties. He qualified feelings, impressions, and emotions that precede the full conscious expression of design features as "Firstness". Of course, Firstness cannot be limited to psychology and biology since it also points to a philosophical category that considers the "indeterminable range of possibilities". But both on the psychological level and on the philosophical level, the situated body experiences **intuition**[27] that opens this realm of possibilities. The challenge for design is how to deal with these impressions, and materialize them into design proposals. While impressions and potentiality are fundamental to the design sensibility, "Secondness" defined by Peirce, that is the actual elements of a situation including humans and non-humans, is what designers have to work with. The composition of the new artifact is one of the stages where design ingredients are tested up against each other. To design and to observe design is to use methods that capture these dynamic tensions and consider the agency of the elements of conception. Another interesting question is how balance, and counterbalance, tension, and proportion, interaction and relation, not only with other human beings but also between material and conceptual components are part of the design process. Finally, "Thirdness", that is the institutional inscription of knowledge and experience is also, in my opinion, this region where design/practice questions cultural and institutional formats, and where design/conception challenges established disciplines.

This very short summary of disciplines that deal with the human from a Foucaldian perspective and the presentation of the semiotic processes based on Peirce and how his categories are relevant to design constitute the theoretical background of my research. From both perspectives, it is obvious that all dimensions of meaning making must be taken into consideration if one wants to understand design practices, processes of interpretation, and to grasp the agency and evolution of forms. How we can engage this theoretical background in design research will be developed in the following chapters. But before we consider the role of social sciences in design, I want to specify what the humanities mean as they are a cornerstone of this text.

1.4 The Humanities and the Designed Object

Many people have asked me why I feel so strongly about the role of humanities in design. This is a fair question that requires several explanations. While in the next two sections, I want to present the humanities in relation to the methodologies and epistemology of design, here I want to focus on the designed object.

---

[27] Petitmengin (2003).

1.4.1 The Extension of the Media Sphere

Today most of our artifacts are hybrids that involve tangible matter and digital matter. This extension of the digital domain is *de facto* an extension of the media sphere. Elaborating on Agre, I feel that the full impact of the computer as a media has not yet been fully acknowledged.

> Computers are representational artifacts, and the people who design them often start by constructing representations of the activities that are found in the sites where they will be used. This is the purpose of systems analysis, for example, and of the systematic mapping of conceptual entities and relationships in the early stages of database design. A computer, then, does not simply have an instrumental use in a given site of practice; the computer is frequently *about* that site in its very design. In this sense computing has been constituted as a kind of imperialism; it aims to reinvent virtually every other site of practice in its own image[28].

First, the overflowing number of communicating devices has made even more urgent the need for a better merging of content-oriented preoccupations with functional focus. Hence the new role of the humanities. We need them to understand how to design intellectual technologies that cannot be considered as mere commodities and utilities but are also media that engage our relationship to a social and cultural project. In his speech at Stanford on June 12, 2005, Steve Jobs recalled the role that calligraphy played for him in the building of the MacIntosh. More than just about typography, this discovery led to a broader conscience of what a computer is. Not solely an object produced from science but an artifact that is related to others by aesthetics and history. An artifact that is about writing and reading, and not solely about processing information.[29]

Even though they are "new media", there seemed to be a conspiracy to wipe out any mention of their materiality from everyday language and even from some socio-

---

[28] Agre (1997).

29 http://news.stanford.edu/news/2005/june15/jobs-061505.htmlReed College at that time offered perhaps the best calligraphy instruction in the country. Throughout the campus every poster, every label on every drawer, was beautifully hand calligraphed. Because I had dropped out and didn't have to take the normal classes, I decided to take a calligraphy class to learn how to do this. I learned about serif and san serif typefaces, about varying the amount of space between different letter combinations, about what makes great typography great. It was beautiful, historical, artistically subtle in a way that science can't capture, and I found it fascinating.

None of this had even a hope of any practical application in my life. But 10 years later, when we were designing the first Macintosh computer, it all came back to me. And we designed it all into the Mac. It was the first computer with beautiful typography. If I had never dropped in on that single course in college, the Mac would have never had multiple typefaces or proportionally spaced fonts. And since Windows just copied the Mac, it's likely that no personal computer would have them. If I had never dropped out, I would have never dropped in on this calligraphy class, and personal computers might not have the wonderful typography that they do. Of course it was impossible to connect the dots looking forward when I was in college. But it was very, very clear looking backwards 10 years later.

logical research on technologies. A quick survey of the vocabulary shows that information technologies seem to be totally immaterial in a number of ways. In the field of computer sciences, one now speaks of "ambient intelligence", "pervasive computing", "ubiquitous computing". In design, one speaks of "interaction design", "service design", "experience design". In social sciences, researchers who are interested in technologies will speak of "socio-techniques". The technology seems to fade in the general landscape or is transparent to its uses. First, the emphasis is on the systematization of "computing logics" that goes with its generalization in places as diverse as home, city, public buildings, etc. Second, the focus is on activities or "experiences" made possible by the technology. In this scenario, researchers do not stop at a clear description of what people have in their hands and how they make sense of it, but concentrate on understanding how this "black box" changes the way we do things, like work, travel, vote, etc.

However, the founding fathers of interaction design argue that they have been working with signs and documents when working with computer interface. This is made very clear by Bill Moggridge[30] who shows how the concept of "document" is not only the final goal of the activity, it is the founding and structural metaphor of the way the computer system works for the user. A number of semiotic properties define the notion of document: first it is a distinct plane of meaning, with a structure and modalities of distribution. It is made of signs that are brought together and constitute a text whose goal is to transmit a message. Second, in terms of use, the document is what makes the application possible. People use the texts to access the functionalities and not vice-versa.

> Like the Star, it was document-based rather than application-based, a regression that was forced on the design of the Macintosh to achieve the reduction in cost. You saw the documents, and the applications were just the things that backed different document types, resulting in a more coherent system than if you had to load applications individually.[31]

Designers need to consider the specificity of their medium not only through the more or less established uses that shape it but also through an understanding of its aesthetic properties. However, Koskinen reports that: "for a variety of reasons, design and media have lived parallel lives rather than mingled seriously".[32] This book is an attempt to weave together these two strands. How do artists, designers, and engineers presuppose different definitions of media? How does their definition impact the way they design them? My assumption is that humanities are then useful because they bring a semiotic contribution to design in addition to the functional and technical points of view.

[30] Moggridge (2007).

[31] Moggridge (2007).

[32] Koskinen et al. (2011).

1.4.2 Design and Media Studies

As pointed out by Kroes, one has to take into consideration the properties of what one designs, to invent and change the attributes of an activity: "an analysis of the design process of technical artifacts should therefore take into account the specific nature of those objects".[33] Design practices are intimately related to what they produce. Within such a perspective, the impact and relevance of the humanities could be limited to designing proper media like television, the internet, etc. However, a considerable branch of design research has focused on how we make sense of things in general and has come up with a communicational model of design whereby the designer communicates to the user through the channel of the artefact. Mostly influenced by Dewey, these researches developed a pragmatism of design which is founded, on the one hand, on the situation and, on the other hand, on the object in situation. In particular, for Crilly, artefacts can be considered as communicative media since people interpret the objects and designers anticipate this interpretation.[34] From the designer, there is an intentional evocation of thoughts, feelings, experiences, or actions, which influences consumers' interpretation. Users infer about the designer's intentions.

In this respect, the humanities are also needed because the discipline studies and compares representations and mediums, from a historical perspective to follow their evolution, and from a contemporary cultural perspective to understand the differences of languages and formats. Therefore, I think their contribution is not only needed because of the increasing number and complexity of computing systems, but also because using the humanities is necessary to understand how people use a variety of languages, material inscriptions, recording tools, and distribution systems to make sense of the world.[35]

1.5 The Humanities as Active Methods for Design

But the humanities are not solely bringing an analysis of iconographic data and tangible artefacts, they also introduce concepts useful for the analysis of design practices. In fact, the book posits that the humanities have a way of looking at creative practices that are useful to invention. Naming, narrating, composing, debating that contribute to radical innovation are all practices that have been thoroughly studied by the humanities.

This book also posits a paradox: materiality is important because no one yet knows what the technologies are going to be for at the time of their invention. While thinking of possible uses then trying to invent new artifacts to support these antici-

---

[33] Kroes (2002).

[34] Crilly et al. (2008).

[35] Mitchell (2002).

pated activities is one way of looking at invention, another is to make a device and then to see how it brings new meanings and activities through the interpretations and performance of the users. The formal shape, the aesthetics of the technology are going to give it meaning without the activity being necessarily well defined in advance. Thus, this book examines how engineers, designers, artists and social scientists strive to come up with methods and tools that they use to bring together cultural and functional meaning despite the absence of formalized goals or perhaps thanks to this indecision.

1.6 The Humanities as Part of the Epistemology of Design

Following this lead, I also think that the humanities are important for the epistemology of design. The easiest way to consider this is that they belong to the orchestra of disciplines working together in design. But more importantly, I contend that the humanities provide the framework that enables the interdisciplinarity or "in-discipline" of design to be considered as a form of "composition" with the different constituents of disciplines (knowledge, models, theories, facts…) that support conception. The in-discipline of design is an aesthetics of design that we need to study from a formal standpoint and not only from a managerial point of view. Design/conception is a collage of elements of disciplines, based on their deconstruction through their under-determination by each other as I will develop in the conclusion. But let me get back to how this realization happened in the field.

1.7 In the Field: Research Through Design

Most of the examples and situations come from my design experience within engineering or artistic teams. My research is a "research through design" as stated by Fallman:

> The interaction design researcher should not be part of the design team as an outside observer, first and foremost a researcher, but rather be part of the design team as a *designer*. The interaction design researcher thus becomes involved in actually putting things together, shaping the form of something new.[36] This process calls for a certain level of participation and commitment on the researcher's part[37]—involvement and participation in a team effort, and a commitment and engagement to build successful products and services—that is unobtainable by an outside observer.[38]

In fact, it is almost impossible not to get involved in a creative situation as, in my experience, every word, reaction, is interpreted as a design cue by the creative team.

[36] Nelson and Stolterman (2002).

[37] Coyne (1995).

[38] Fallman (2008).

Being part of design situations therefore meant that either I would minimize that impact, or that I would allow myself to contribute and see how the team members reacted to my suggestions or whether they required different inputs – which is what I finally opted for.

1.7.1 Research Through a Collection of Breaching Experiments

In the field, the humanities were helpful as they helped me concentrate on breakthroughs by dealing with their aesthetics (new style of painting, new literary genre, original musical interpretation). I considered the "ultimate particular", which meant comparing regularities, remediations, reformulations and broadly speaking redesign to spot the unusual, the strange, the original. I think that this focus on cultural innovation, that tracks novelty and originality against convention and tradition, offers ways of understanding the invention and development of new media. To do so I had to build and analyze corpuses of objects. While a common method in the humanities, Bill Gaver gives some clues on how to proceed from an HCI (Human Computer Interaction) and design perspective:

> One of the valuable roles of design theory, from this perspective, is in making accessible the kinds of decisions and rationales that comprise an artifact's embodied theory, or give dimensionality to its design space. In this case, however, then instead of theories predominating, with design examples serving as mere illustrations, design theory is best considered a form of annotation, serving to explain and point to features of 'ultimate particulars', the truths of design.[39]

Therefore, design examples are not illustrations of theories but rather try to make accessible the rationale of design through its making and its materiality.

I also want to explain why my selection includes not only objects produced by designers or engineers but many works of art. I am aware that the latter are a rather unusual occurrence in a book about design. I have previously explained that I want to see common points between the different types of productions. I also find that works of art often display striking features that question how we look at things. Literature or plastic art, open a time and space where spectators can question everyday language and representation. In other words, they not only engage spectators in environments that question the everyday experience, they also open a space of reflexivity. They also generate new methods to interpret different media and how spectators relate to them. This is a second important aspect. As Clément[40] or Aarseth[41] remark, contemporary art does not rely solely on the gaze. Installations, performances, as well as interactive works of art cannot solely be looked at and

[39] Gaver (2012).

[40] Clément (2000–2001).

[41] Aarseth (1997).

interpreted. They have to be played and performed. The scholarship that art experts bring into the analysis of art pieces has evolved from issues of narrative, themes, montages, materials, etc. to include a careful analysis of how people are directed to perform the art piece. Art critic meets ethnographer as she has to experience the piece herself and observe how other people use it. The practices and interactions are then also considered as form. I therefore look at works of art as breaching experiments – that is unusual events or artifacts, consciously fabricated, and that disrupt the usual course of affairs to better understand innovative conception.[42] Indeed, the concept and method of breaching experiment elaborated in ethnomethodology demonstrates how human groups structure themselves around explicit values and modes of organization after they have been disturbed. However, I am more interested in the formal aspect of these breaching experiments (a point that is not very much developed in ethnomethodology) rather than their impact. Design and art are a case in point since working on and creating a new object necessarily challenge cognitive, aesthetic, and social expectations. What are then the properties and agency of objects in a design situation? How do they push us to reorganize our lives? How do we push back, or make sense of them? These questions are central in this book. Using research through design, I want to see how designers, artists, and engineers play and create new meaning through new tangible artifacts.

1.7.2 Reflective Research Practice: The Two "Moments" of Research Through Design

Classical ethnography (for example, Goffman[43]) outlines two moments in ethnographic research: the moment of complete immersion in the field, and the moment of reflection. As previously mentioned, I was deeply involved in the process of technical development within projects, helping to shape the form of the different versions of the invention, from discourses in research grant application to the demonstrators. After accumulating field notes and a lived experience of these breaching experiences, I enlisted many friends' and colleagues' assistance to sift through this data, thus beginning the second ethnographic moment. The distance and perspective that my colleagues provided helped to more rapidly make sense of what had been a very deep and active engagement with the process of technical development. But there is more to it than a simple chronology of research steps.

While I describe design situations, the reader will see that I also reflect on my participation in the process. Or course, I am sensitive to Button's admonition to remain conscious of "what people have to know, and how that knowledge is deployed in the ordering and organization of their work".[44] Suchman's ethnometh-

[42] Garfinkel (1991).

[43] Goffman (1959).

[44] Button (2000).

odological studies of such situated actions in the corporate sector have laid the groundwork for this kind of reflective practice in HCI.[45] However, beyond the usual attention for the limitation of a situated research, I think that the strong expression of a set of beliefs and knowledge elements is part of the creative process itself. Indeed, to structure this reflexivity but also the invention, most of the projects were built around partnerships that allowed at least three different viewpoints (engineering, communication, design) that shed very different lights on the technical object and in fact revealed different underlying conceptions of what a technology is. In other words, I represented one more voice that could under-determine and also be destabilized by the other disciplines. The reflection on my disciplinary standpoint is therefore less a methodological caution than a way to question the possible articulations between disciplines. Part of my research effort went into testing the relevance of a partnership between researchers in design and in information and communication science and in engineering. In every program, I had "insiders' feedback" and requirements from the partners in the project with whom I would discuss my contribution to the project, and "outsiders' feedback", from other social and human scientists who would review this fieldwork with me.[46] In this context, I believe that information and communication sciences have a dual role to play. They analyze epistemic communities not only in terms of the players involved and the relationships between these actors, but also in terms of specific forms of communication: situations, objects, texts, documents, media. Second, they can contrast their concepts of representation and knowledge with those of computer scientists, thereby provoking interesting developments in the design process.

1.8 Book Overview

In addition to this introductory chapter, there are six chapters that deal with the elements and operations of design. The first two chapters discuss the object and subject of design. The third deals with the languages of design. Chapters 5 and 6 tackle the issues of composition and expansive conversation. I finish with the epistemological in-discipline of design. Here is a summary of the different chapters and of the main technologies that are discussed.

The situation of the Codesign lab in Telecom Paristech made it easier to discover and participate in inventions related to the field of information and communication technologies. However, the examples are also based on other areas of research and development in particular health and biology thanks to the discussions with critical designers like James Auger and Anab Jain, collaborative work with Anne-Françoise Schmid (philosopher of sciences) and Muriel Mambrini-Doudet (former head of the

[45] Suchman (1987).

[46] Here I would like to thank Alison Powell, Aude Guyot, Camille Jutant, Mathias Béjean, Tiphaine Kazi-Tani, and Cédric Mivieille who helped me build a proper theoretical framework.

National Institute of Agronomical Research in France) and the doctoral work of the designer Max Mollon on neurodegenerative diseases.

In Chap. 2, I mostly study interactive and reflective systems that give us an understanding of our relationship to computing systems through image (Agnes de Cayeux and Antoine Schmid's artworks) and through sound (François Pachet's "Continuator"). These representative systems capture and give new meaning to our gestures, hence elaborating a new holistic relation with smart machines. The Second Chapter offers to switch from the interactive to the reflective design paradigm. The book posits that interactive design as a paradigm of design relies on a model that focuses on activities and does not question the values, the aesthetics of the artifact and therefore restricts the expansion into new norms, or new aesthetics. On the contrary, reflective design is about methods that generate new questions about functionalities, but also personal emotions, social rituals, formats, and knowledge. Amongst tools and methods of reflective design are information technologies that are defined not only in terms of activities but also as semiotic and meaning-making technologies that open new spaces for representation and reflection. These technologies can support a design practice that consists of stepping back to gain a different perspective, looking at alternative options, and discovering the embedded values. Here information and communication technologies (ICT) are not only considered as a field that is being shaped by multiple actors, but as design tools, and somehow a metaphor for the design goal: to "denaturalize" activities and formats to be able to lay new foundations for a design project. There is, therefore, a double need to understand the properties of these reflective artifacts: first to understand these technologies to better design them, second to understand how they offer a design space with tools and metaphors to think about design and conception properly. More generally speaking, media offer an autonomous and reflexive material that help designers invent new aesthetics.

In the Third Chapter, I contrast the sociology of use and the engineering research perspectives on the "user". Both disciplines are major stakeholders in defining how people interact with computing systems. I want to show that their perspectives can be augmented by the field of aesthetics that concentrates on the sensitive experience of the object and the field of media studies that focuses on audiences and spectators to understand how they engage their literacies to produce meaning. Designers of computing systems have to go beyond a simple model of activity to include the aesthetic experience and interpretative faculties of people that have to read and write these technologies. I will therefore show how the design process includes several "figures" of the user. These figures are poetic productions, and indirect representations of different models of real users: one who manipulates technologies to act, as well as an aesthete and a reflective individual. The objective is to obtain a richer view of the "users" and to engage in a debate on how they form a complex system with our objects. However, the chapter also challenges the idea of user-centered design by pointing out that it could be more important to confront "figures" of the users to support the expansion of the design project beyond a mere replication of standard uses. In other words, the book posits that the non-unified figures of the user are a lever to actually invent new artifacts. Ethnological approaches are therefore not directly but indirectly useful for design because they give material to imagine potential future uses. In Chap. 3, I analyze a pervasive game based on a distributed architecture and RFID tags that we developed for Paris Museum of Arts and Crafts. This experience shows how mobility is now part of the reflection about accessibility of culture but also how pervasive computing systems redefine space and contents as well as the activity in the museum and its visitors.

The Fourth Chapter introduces a debate between the sociology of science and the humanities. The sociology of science has demonstrated the role of rhetoric in the making of science. I want to show that certain productions have other qualities – poetic qualities – that are best understood with the tools of the humanities. In particular, literary studies examine the way authors structure narratives to build fictions that make sense for the readers because they touch some profound experience of the world. This chapter aims at exposing how research is also the place for a poetics of science. The objective is for inventors to be able to claim and develop these practices that contribute to the maturity of the invention. In this fourth chapter, engineers emerge as poets and research is just as much about naming things as inventing things. Researchers work on words, acronyms, expressions, to give an identity to their production. In a globalized field of research, it is also a linguistic experience of translation. Finally doing research needs narrative skills. The researched technology is always the hero of complex stories that weave the present and the future together. Chapter 4 considers several technologies like Bluetooth and Virtual Intelligent Agents, as well as research fields like Artificial Intelligence. What is interesting about these examples is that they challenge technical assumptions but more importantly they challenge the assumed identity of objects. It is particularly striking with Anab Jaïn's work on visually impaired people and new sensitive systems.

In Chap. 5, I look at design as a practice of composition in a field of tensions. By laying out materials, ideas, forms, models of communication and activities, designers organize their practice not so much as a sequence of events but as a field to compose within. Rather than using the metaphor of the project, I look at the metaphor of the matrix. The matrix contradicts the model of design thinking that starts with an abstract idea that is gradually implemented through an iterative process. This classical model dating back to Aristotle's vision of architecture, does not account for the fact that design/conception and design/practice are bringing together materials in unexpected ways without necessarily following a defined plan. However, I find interesting to analyze how composition is a projective abductive practice. To do so, I first study the design of an e-learning platform called VUE. Before the MOOC trend, many academic and industrial actors tried to offer reliable and attractive online training. In this example, different tools are used that tackle the main difficulties like sharing, evaluating, but also creating a group that constitutes a real community of learners and doing so on multiple media (phone, tablet, computer). This example is useful in that it shows the contrasting semiotic practices of the designers. I then look into the project of an interactive multimedia theater show using three stages in three different cities, exploring another form of long distance presence. Beyond the artistic ambition, the whole writing project was an exploration of different types of writing software, each unveiling different aspects of what a show can be. Both projects show the challenges of composition.

The Sixth Chapter meets again with the sociology of science and confronts it with political philosophy – in particular Habermas's theory of communication – and critical design. From the point of view of design, the question is how stakeholders organize a debate around their production and how it sustains the generativity of the design process. In this chapter, I analyze how researchers and creators structure generative debates with specific situations of communication and objects. In between art and research there is a variety of means to trigger then build a constructive dialogue. The objective is to help inventors recognize and include these elements of dialogue with a larger audience to fully endorse the political dimension of their work. This leads us to consider a turning point in design activities: designers produce not only objects, but they produce "things" whose identities are in question, hence the need for generative debates that contribute to the invention. Chapter 6 examines three examples that shape the way artists, designers and researchers challenge their own perception and that of their users and audiences. The first is about the circulation of information on internet and how the media can be orchestrated to build a public opinion. The second example takes scientific demonstrators of Adhoc networks and how information is distributed between screens and objects leaving some leeway for the active performance and interpretation of the users. Finally, the third example is taken from microbiology, to examine the range of possible uses for the cultivation of human tissues and how we can debate about science and ethics.

Chapter 7 concludes this book by reflecting on the "in-discipline" of design. Designers claim that their practices are transversal, multidisciplinary, and holistic. I suggest that this claim is unreachable if one does not have a precise idea of what pluridisciplinarity means for the design of new objects. While pluridisciplinarity is the underlying thread of all the chapters in this book, the last chapter tries to give it an epistemological definition. How can disciplines work together as well as put together different forms of creativity from design and art to social science, humanities, and engineering sciences? Elaborating on Françoise Schmid's concept of "integrative thing", I want to show how through design/practice, disciplines under-determine each other, leaving space for a radical unknown to emerge. Design is therefore no longer a Leonardesque fantasy of mastering all the known disciplines, but rather the dynamic activity that launches concepts, facts, methods, between disciplines so that they can come up with new concepts and artifacts, or situations. In this chapter, I look again at the pervasive game project in the Museum of Arts and Crafts to show that design/conception is a deconstruction of disciplines as such. The emerging technology is made of heterogeneous elements that a single discipline cannot synthesize.

References

Aarseth, E. J. (1997). *Cybertext: Perspectives on ergodic literature*. Baltimore: The Johns Hopkins University Press.

Agre, P. (1997). Toward a critical technical practice: Lessons learned in trying to reform AI. In G. Bowker, L. Gasser, L. Star, & B. Turner (Eds.), *Bridging the great divide: Social science, technical systems, and cooperative work*. Hillsdale: Erlbaum.

Akrich, M. (1990). De la sociologie des techniques à une sociologie des usages. *Techniques & Culture, 16*, 83–110.

Akrich, M. (1992). The de-scription of technical objects. In W. Bijker & J. Law (Eds.), *Shaping technology/building society. Studies in sociotechnical change* (pp. 205–224). Cambridge, MA: MIT Press.

Benjamin, W. (1935–2010). *The work of art in the age of mechanical reproduction*. Createspace independent publishing platform.

Buchanan, R. (2001). Design research and the new learning. *Design Issues, 17*(4), 3–23.

Button, G. (2000). The ethnographic tradition and design. *Design Studies, 21*(4), 319–332.

Chow, R., & Jonas, W. (2010). Case transfer: A design approach by artifacts and projection. *Design Issues, 26*(4), 9–19.

Clément, J. (2000–2001) Hypertextes et mondes fictionnels (ou l'avenir de la narration dans le cyberespace). *Ecarts, 2*.

Coyne, R. (1995). *Designing information technology in the postmodern age: From method to metaphor*. Cambridge, MA: The MIT Press.

Crilly, N., et al. (2008). Design as communication: Exploring the validity and utility of relating intention to interpretation. *Design Studies, 29*(5), 425–457.

Cross, N. (2011). *Design thinking: Understanding how designers think and work*. Oxford/New York: Bloomsbury Academic.

Davallon, J. (2004). Objet concret, objet scientifique, objet de recherche, *Hermès (Paris. 1988), 2004, 38, fascicule thématique Les sciences de l'information et de la communication: savoirs et pouvoirs*, http://documents.irevues.inist.fr/handle/2042/9421

Dow, S., Wendy, J., & Mackay, W. (2013). Projection, place, and point-of-view in research through design. In*The sage handbook of digital technology research* (pp. 266–285). Thousand Oaks: SAGE.

Fallman, D. (2008). The interaction design research triangle of design practice, design studies, and design exploration. *Design Issues, 24*(3), 4–18.

Findeli, A., & Coste, A. (2007). De la recherche-création à la recherche-projet : un cadre théorique et méthodologique pour la recherché architecturale. Nantes: *Lieux communs. Les cahiers du LAUA, 10*, 139–161.

Foucault, M. (1966–1994). *The order of things: An archaeology of the human sciences* (reissueth ed.). New York: Vintage.

Garfinkel, H. (1991). *Studies in ethnomethodology* (1st ed.). Cambridge, MA: Polity Press.

Gaver, W. (2012). What should we expect from research through design? In *Proceedings of the SIGCHI conference on human factors in computing systems*, 937–946, http://dl.acm.org/citation.cfm?id=2208538

Gentes, A. (1996). *La communication publique : de la mise en scène à la stratégie, de la norme à la démocratie*. PhD thesis under the supervision of Professor Bernard Miège, Université Grenoble 3. (not published).

Goffman, E. (1959). *The presentation of self in everyday life* (1st ed.). New York: Anchor.

Hatchuel, A., Weil, B., & Masson, P. (2012). Towards an ontology of design: lessons from C–K design theory and Forcing. *Research in Engineering Design, 24*(2), 147–163. https://doi.org/10.1007/s00163-012-0144-y.

Jutant, C., Gentes, A., Béjean, M., & Mivielle, C. (2013, August). *Design, meaning making and constructive fixation Conceptualizing semiotic conditions to the process of designing* (pp. 3509–3519). Tokyo: IASDR.

Koskinen, I. K., et al. (2011). *Design research through practice: From the lab, field, and showroom*. Burlington: Morgan Kaufmann Publishers In.
Kroes, P. (2002). Design methodology and the nature of technical artefacts. *Design Studies, 23*(3), 287–302.
Latour, B. (1987). *Science in action: How to follow scientists and engineers through society*. Cambridge, MA: Harvard University.
Leroi-Gourhan, A., & White, R. (1993). In A. B. Berger (Ed.), *Gesture and speech*. Cambridge, MA: The MIT Press.
Manovich, L. (2002). *The language of new media*. Cambridge, MA: The MIT.
Mitchell, W. J. T. (2002). Showing seeing: A critique of visual culture. *Journal of Visual Culture, 1*(2), 165–181. https://doi.org/10.1177/147041290200100202.
Moggridge, B. (2007). *Designing interactions* (1st ed.p. 81). Cambridge, MA: The MIT Press.
Nelson, H. G., & Stolterman, E. (2002). *The design way* (p. 327). Englewood Cliffs: Educational Technology Publications.
Nelson, H. G., & Stolterman, E. (2012). *The design way: Intentional change in an unpredictable world* (2nd ed.). Cambridge, MA: The MIT Press.
Norman, D. A. (2002). *The design of everyday things*. New York: Basic Books.
Petitmengin, C. (2003). *L'expérience intuitive*. Paris: Editions L'Harmattan.
Shank, G. (2001). It's logic in practice, my dear Watson: An imaginary memoir from beyond the grave, *Forum Qualitative Sozialforschung/Forum: Qualitative Social Research*, *2*, 1. http://www.qualitative-research.net/index.php/fqs/article/viewArticle/970
Simon, H. A. (1996). *The sciences of the artificial* (3rd ed.). Cambridge, MA: The MIT Press.
Simondon, G., & Hart, J. (2001). *Du mode d'existence des objets techniques*. Aubier.
Souchier, E., Jeanneret, Y., & Le Marec, J. (2003). *Lire, écrire, récrire : Objets, signes et pratiques des médias informatisés*. Paris: Bibliothèque Publique d'Information.
Suchman, L. A. (1987). *Plans and situated actions: The problem of human-machine communication* (2nd ed.). Cambridge, MA: Cambridge University Press.

Chapter 2
From Interactive Design to Reflective Design

2.1 When Objects "Talk Back": Design as a Strategy in Critical Aesthetics

As regular users of the internet, we have expectations about the interactivity of a system. But works of art always seem to destabilize these pre-conceptions. In the early 2000s, as I was teaching new media art, I worked 2 years in a row with a group of about 20 students to understand what was so special about these art pieces that so contradicted our newly acquired "reflexes". At that time net art was burgeoning[1,2] and students repeatedly felt "manipulated" because operations such as pointing, opening a window, rolling a menu were challenged. Digital artifacts seemed to work like distorting mirrors, sending back a reflection of our activities that was both true and deceitful, translating and interpreting our gestures and commands. The slight distortions between what we anticipated and what we actually got, the fact that our activities were staged or responded to in unexpected ways, all this forced the group to step back and think about the specificity of this aesthetic experience. The works of art actually taught us that what we took for granted – for example, the real time response of a system, as well as a certain number of semiotic traits of the interface – were cultural constructs that served a vision of fluidity in the man-machine system. The concept of reflectivity finally emerged alongside my collaboration with a number of artists exploring computers and the internet as their main materials. While working with them I realized that I needed a better understanding of how these technologies were designed. In particular, I felt that the expression "interactive" design did not quite capture the work that designers or artists did with the aesthetics of these machines and their role in society. Looking at digital art works was therefore not only discovering a new branch of art, it actually worked as a starting point

[1] Greene (2004).

[2] Bosma (2011).

A. Gentes, *The In-Discipline of Design*, Design Research Foundations,
DOI 10.1007/978-3-319-65984-8_2

for a journey to their creation. Studying new media art was a way to get a sense of how these machines could be designed differently.

The whole experience was a lesson not only about art but about design in general. Designing means inventing new gestures, interfaces, services, technologies. Users' assumptions are therefore challenged. Through design, what we think of as "natural" is deconstructed as truly cultural and therefore changeable. The whole book is about what it means to organize a process that denaturalizes our relation to already existing artifacts. The focus of this chapter is that certain objects can foreground this questioning, and help shape the designer's and users' reflection. I want to show that certain objects "that talk back" are part of a reflexive design strategy for both designer and user. I therefore want to look at some objects that were designed so as to question their own modality of existence. This particular selection gets us closer to what design research calls "critical reflection" as an important principle of design. I think it will help us move from design, seen as the best adjustment of features to functions, to design considered as a deliberate plane for representation and reflection that therefore needs not only engineering or social sciences skills, but the humanities.

2.1.1 "The Reflective Practitioners"

In design, the expression "reflective practitioner" was introduced in the '80s by Schön, when analyzing the iterative process of reflection going on when designing artifacts or doing architecture. It has been used in a more systematic way by Sengers et alii as a way to avoid the limitations of one set of practices. It is, therefore, associated with critical reflection: "Critical reflection on the limitations of the field's methods and metaphors can help us to see the world in a new way, identifying and weighing new technical possibilities".[3] Critical here is considered as a way to expand the range and scope of the design space by de-naturalizing the production and effects of artifacts. It goes against the more traditional vision of design as a series of necessary steps towards a limited amount of options. On the contrary, critical designers want to qualify the results, to evaluate them in terms of social and cultural values to avoid the evidence of well-used design metaphors. As a result of this critical enquiry, "choice" between various value options can be reintroduced in the design process. "We define **'reflection'** as referring to critical **reflection,** or **bringing unconscious aspects of experience to conscious awareness, thereby making them available for conscious choice**[4]". The authors point to a number of practices that introduce reflective design. First, participative design introduces some divergence by organizing a participation of the public in the production of the artifact or service. The diversity of opinions is, therefore, supported by a diversity of actors. But they emphasize the fact that the actors – designers and potential

[3] Sengers et al. (2005).

[4] Ibid.

users – share the opinion that their goal is worthy of design and even if different means or interpretations may clash during the project, the final end is not questioned. This does not allow for the definition of altogether different social goals, in other words, it does not facilitate radical innovation. The authors also mention "Value sensitive design" that focuses on values behind design projects. Here it is the process that is central to supporting a diversity of points of view. Going a bit further, "critical design" as defined primarily by Dunne and Raby at the Royal College of Art in London, actually challenges the goals. Their work is a constant questioning of the ethical implications of major and diverse technological and scientific trends such a DNA testing, nuclear power, mass surveillance technologies, etc. They build objects of speculative design that embed unsettling scenarios based on what could be done with these technologies. They are very much focused on issues of reception amongst the audience. The main difficulty in their design practice is to strike a fine balance between the canny and uncanny aspects of their artifacts so that they avoid triggering a pure and simple rejection from the audience.[5] I will further discuss design as debate in Chap. 6 but I want now to focus on objects that are experienced as reflective and expansive to understand what their properties are, and how they help the designers to step back from everyday assumptions.

2.1.2 *Outline of the Chapter*

When Schön says that the situation talks back, it is because certain arrangements do not feel "right" to the designers. Elaborating on this, I want to point that artifacts can be designed with certain formal traits that prevent them from fading in the fabric of everyday life and thereby provoke this reflectivity. To understand what is meant by reflective artifacts I want to look at a few examples that display this critical stance.

I take my first example from Human Computer Interaction (HCI). The example – François Pachet's "Continuator" – shows an attempt to suspend our musical activity and open a space of dialog that might contribute to some reflection from its users. Two artistic examples go much further on the critical scale but also strive to question our relationship to machines. Finally, I describe two research use cases where a "bug" helped us move forward in our design project. The latter examples, taken from research projects in distributed systems, show how representation supports the necessary distance that designers need to pursue, so as to extricate themselves from existing artifacts and everyday assumptions.

The following part of the chapter is devoted to positioning the paradigm of reflectivity in design with regard to the paradigm of interactivity. Both semantic fields presuppose a feedback loop between user and system, but I will argue that introducing the concept of reflective technologies helps us think about design as a critical activity that expands the design space. The goal of this chapter is to show that information technologies can be described as reflective technologies. In that

[5] Gentes and Mollon (2015).

respect they are more than just an application field of design, they are a tool for designers, and they are a metaphor for the design space in so far as it has to be both a reacting and reflective space. The last sections of this chapter present how humanities have studied the question of reflectivity and how these analyses can help us better understand the process of conception. I also look at some particular formal strategies that tend to give a special status to the object so that it is experienced as a stepping-stone for new design.

2.2 Sound "Mirror": A Theory of the Mirroring Effect

I will start with an exemplary system, "The Continuator" (Fig. 2.1), developed by François Pachet at the Sony Lab in Paris. Pachet coined the term "reflective interaction" for his musical system, a piano that borrows from the musician/user's style to play a sequence on its own. Pachet used the concept of "reflective interaction" to differentiate his system from "well" designed HCI whose goals are to be so seamlessly integrated into our actions that we do not have to think about them as they disappear behind the task. But, as a musician and specialist of digital media, he also wanted to explore how to open a rich space of signs and references that also reflect on our activities. He therefore invented musical tools that can reflect musicians' practices.

Fig. 2.1 Pachet "The Continuator" (2006b)

2.2.1 Interactive Systems and Pleasure

The highly significant hypothesis developed by François Pachet, in "the Continuator" is that we can develop an emotional relationship with music by receiving a mimetic feedback from our own sound production. François Pachet has therefore designed a musical interactive and reflective system that creates a desire to play:

> …More precisely, we propose to consider the class of interactive systems in which users can interact with virtual copies of themselves, or at least agents that have a mimetic capacity and can evolve in an organic fashion. To make this imitation efficient, there are a number of characteristics that we consider important to define reflexivity in interactive systems.[6]

Among the most basic elements of this experience, François Pachet examines what he terms the "mirroring effect":

> Similarity or mirroring effect. What the system produces sounds like what the user himself is able to produce. This similarity must be easily recognizable by the user, who must experience the sensation of interacting with a copy of herself. Similarity is not equivalent to mirroring. For instance, a systematic echo or repetition of the phrases played by the user does not induce such a sensation.[7]

A musician plays notes on a piano. These notes are interpreted by a program which then presents follow up musical notes. "Interactions with the users are analyzed by IRMS to build progressively a model of this user in a given domain (such as musical performance). The output of an IRMS is a reflexive *mimetic response* to a user interaction[8]". The system has been tested with experienced composers as well as children. All testers recognized their own personal "style" in the suggestions by the system. In other words, what captivates the users and what touches them in these interactive systems is the personal yet unusual aspect of what they discover about themselves, the instrument, and, possibly, music.

The issue of similarity is important: the machine is not programed to produce an exact copy of what the player plays. It creates nuances that increase the complexity of the relationship so that the player not only recognizes herself but also has an experience of discovery. The *Continuator* is indeed a mirror, but an enlarging one exposing interpretive opportunities in interaction. The passage from interactive to reflective is supported by an important notion: the mimesis. This concept inherited from thousands of years of painting as I detail later, captures the likeliness but also the distortion that is at the heart of the artistic process.

The art historian Phay-Vakalis' analysis of the relationship between painting and a mirror better explains what is at work, not only visually but more broadly in this type of sound, texture and image representation.[9] She takes Manet's famous work, *Un Bar aux Folies Bergères* (1881–1882) as an example. This painting conveys ideas from the early period of photography in which the established myth that paint-

[6] Pachet (2006a).

[7] Pachet (2006a)

[8] http://www.mirorproject.eu/default.aspx

[9] Phay-Vakalis (2001).

ing must imitate nature had lost its hold. In his painting, Manet portrays impossible reflections in the mirror behind the female seller. She appears to be looking at us but, when we examine her back reflected in the mirror, we see that she is looking at a client outside of our field of vision. Gradually the mirror is no longer a trick of the eye but causes us to question what we ourselves are seeing. In the history of art, the use of mirror is always emblematic of an ongoing debate about the role of painting and its "imitation of nature" but also about the place of the spectator when he/she looks at the painting. During the twentieth century, this question was particularly challenging as photography seemed like a much more efficient way to replicate the world. Eventually, mirrors were also embedded in the canvas. The use of actual mirrors inside the painting allows the viewer to enter into the work itself like in Jacques Monory's 1968 painting *Meurtre* (oil on canvas, *bullet impact on mirror*).

> Since a painting can never match photography exactly, the mirror takes the place of the canvas, and becomes a support and surface. No longer a metaphor of mimesis, but a real work of art, [the mirror] is therefore transformed into an "artistic material" and is gradually confirmed as an "operator of exchange" between the viewers and their environment.[10]

The reflective translation of our input by the machine surprises us because it is partly true to what we have produced but offers us a new vision of ourselves. It is therefore not a system designed to test our abilities but it is a system to create exploratory environments of ourselves and others. In other words, sensor systems are not only recording and measuring tools; they are perceived as systems that can introduce different forms and proposals that can intrigue, worry, surprise, or even delight by taking and modifying the user's input and returning a "result" that is both loyal to the source and different from what the user expects.

2.2.2 Pleasure to Learn

The second phase of François Pachet's project shows an interesting turning point. He goes on to demonstrate that mimetic systems are not only pleasurable, they also support self-learning. People learn about a new system but, more importantly, about themselves.

The European project: "MIROR" (Musical Interaction Relying On Reflexion) which was a prolongation of "the Continuator" delved into the important differences between traditional approaches to man-machine interactions that consist of designing algorithms and interfaces that help the user solve a given, predefined task, and the IRMS (Interactive Reflexive Musical Systems) that are designed without a specific task in mind, but rather as an intelligent exploratory space. The principal output of these systems is that, thanks to the "chameleon effect" analyzed by Chartrand and Bargh,[11] users learn about themselves, music, and their relation to

[10] Phay-Vakalis (2001, P. 133).

[11] Chartrand and Bargh (1999).

music. There are educational benefits of such a mirroring system, in particular because mimetism is an essential paradigm of identity building.

This point was also made by Sherry Turkle[12] in her book on virtual identity. She showed that people using "Second World" were not only playing and discovering the potentials of a new technological system but also exploring new facets of their personality and therefore learning about themselves. For example, a man could try himself as a woman through his avatar, thereby exploring the more "feminine" traits of his personality. Even if the virtual character appeared very different from their real self, the interviewees declared a strong likeness to their virtual character that allowed them to gain meaningful insight for their life.

Whether in virtual worlds, or with such systems as "the Continuator", what is at stake is how such systems give users specific means to distance themselves from their everyday selves and usual social or technical interactions so that they can eventually consider other options. Representation and distance offer a new way to learn about oneself, by trying out new roles, new gestures. The reflexive activity is also related to learning by Schön when he analyzes the designer's practice:

> A practitioner's reflection can serve as a corrective to overlearning. Through reflection, he can surface and criticize the tacit understandings that have grown up around the repetitive experiences of a specialized practice, and can make new sense of the situations of uncertainty or uniqueness which he may allow himself to practice[13].

According to Schön, the reflexive practitioner organizes situation of uncertainty and uniqueness. With Wiggins, he notices the primordial role of using media to do so:

> A designer sees, moves and sees again. Working in some visual medium - drawing, in our examples - the designer sees what is 'there' in some representation of a site, draws in relation to it, and sees what has been drawn, thereby informing further designing.[14]

While there are other ways to plan reflective practices systematically - like organizational procedures, or tools and situations of communication with other stakeholders - the use of a media introduces the necessary distance to explore alternatives by sketching and more generally producing a simulation. In "the Continuator", the strategy relies on the magic of the machine working on its own while we stop using it. The practice is reflexive first because it is based on time lags that generate a new kind of listening: the players do not listen for their bugs or accomplishments but explore the musical design space with the machine; second, because there is a discrepancy between the human input and the productions of sounds that suggests alternatives. While I do not know what the kids or the professional musicians learned from the situation, they had a feedback that they could elaborate on whether to learn about themselves and try new activities, or to produce new designs.[15]

---

[12] Turkle (2005).

[13] Schon (1984).

[14] Schon and Wiggins (1992).

[15] Johnston et al. (2005).

Of course, depending on the media, this operation of translation can take several forms: the musical production is imported by the system and changed along some musical rules of composition, or the image of the spectator through her gestures is incorporated in a new setting and disturbed from its usual course. In any event, reflective systems that open a space of representation are not limited to providing the best features to successfully support functionalities. While tools and more generally technologies are geared towards activities and evaluated for their capacity to do so with maximum efficiency, reflexive technologies can hold together both the activity and a reflection about the activity which makes them extremely precious as simulation tools as we will see in Chap. 5. Designing for reflexivity with these technologies to both learn about the assumptions and values embedded in systems and what we expect from them is key to expanding a situation and consciously examining design decisions.

2.3 The Concept of Interactivity: Tricks of a Concept

I find the example of the Continuator very interesting because it is both an interactive and a reflexive system and it tries to derive the best from both paradigms. I will first dwell on what is meant by interactivity and suggest that the metaphor can be misleading. For François Pachet, as well as most HCI designers, what qualifies these technologies best is that they respond to us, they "interact". But as we have seen, this is not only a question of "responding". There are deliberate choices that embed a role of what our interactions with machines should be.

Meanwhile, the concept of interactivity primarily focuses on users involved in an activity and a machine providing the functionality. Software designer and programmer, pioneer of interaction design, Alan Cooper gives a very precise definition: "Almost all interaction design refers to the selection of behavior, function, and information and their presentation to users[16]". Interaction designers focus on the user's goals. In all of Cooper's examples, the idea is that there is a task at hand and that the system must not fail this task: "goals are the reason why we perform tasks".[17] The focus of interactivity is on functions and actions of things and people. Ideally, technologies should be transparent artifacts that disappear in favor of the user's activity.

But some social scientists criticize the expression because it lets people think that there is an "interaction" that may be construed as almost human or sufficiently close to human behavior and therefore imbued with the same qualities, properties, and responsibilities. The media scientists, Emmanuel Souchier and Yves Jeanneret who wrote at the beginning of the expansion of multimedia connected systems in 1999 were particularly concerned that:

[16] Simon (1996, P. 22).

[17] Cooper (2004, P. 29).

> The rhetorical effectiveness of the term interaction is indeed in its ambiguity, which refers to the technical properties on the one hand and the notion of strictly human action of the other. But the action is not just a physical act - which is dealt with by ergonomics - it is also a display of energy endowed with meaning by a subject in a social, historical and cultural context - hence the action is the object of semiology, sociology and philosophy. It is therefore understandable that there is no possible interaction between man and machine. It is therefore appropriate to wonder why the designers of these devices talk of interactivity and why they try to make us think, that the machine has a capacity for action, commitment, an ability to make sense and thus an ability to access to culture as the term suggests[18].

In this text, the authors criticize the word "interactivity" because they consider that a blurring of boundaries occurs that obscures the analysis of what these technologies are. They argue that using the expressions interaction, interactivity, and interactive, while probably saying something about these technologies, also tends to hide the fact that they are artificial machines that do not think. They also imply that to compare human beings to mechanical artifacts is misplaced. Along with critical theory, and in particular Barthes,[19] they suspect that this metaphor naturalizes some ideology or political agenda that hides behind the false obviousness of the expression but also helps with its success.[20]

A quick survey of the history of the expression and its circulation, supports the critical stance by showing how two unrelated fields: physics on the one hand, and social sciences on the other, use these expressions with very different meanings, and how finally they are going to merge in the field of computer systems. The Oxford Dictionary[21] shows an exceptional circulation of the word interaction as early as 1832 (and it appears in the French Dictionary the Littré in 1876). The term at that time refers to a system of elements where the action of each affects all:

- 1832 Taylor Saturday Evening 1833 The infinite excellence ... comprising Interactive Causes which must have products possessing absolutely no affinity with anything exterior to itself.
- 1879 H.W. Warren Recr. Astron xii 257 Yet its interactive atoms can give 400 millions of light-waves a second.

The history of this term shows the mechanical and dynamic aspects as highlighted in the Cultural Dictionary of Science directed by Witkowski[22]:

> The idea of mutual interaction between the bodies or interaction was formalized in classical physics by the concept of force, first in the case of gravitation and then in electromagnetism: one speaks of electric or gravitational forces. With the advent of quantum theory, the notion of force is losing its importance in favor of more abstract theoretical entities (potential) and the idea of action (possibly remote) is replaced by exchange of mediating particles. Thus the word interaction, conveniently vague, has come back to designate the various basic types of mutual actions between physical bodies.

---

[18] Jeanneret and Souchier (1999).

[19] Witkowski (2001).

[20] Marec (2001).

[21] Simpson and Weiner (2002).

[22] Witkowski (2001).

Much later, the word interaction will be used in social sciences. As specified in the Dictionary of notions of the Encyclopaedia Universalis,[23] the concept issued from German philosophy is generalized by the Chicago School. Social psychology describes the rules that organize human relations and their social stakes. Thus, from the '70s, the term tends to encompass human activities and not just the physical or chemical properties of elements.

Since the '70s, the expressions interaction and interactivity have been used and disseminated to describe computing systems. It is worth emphasizing that the terminology that was developed in the context of electricity and electromagnetism is replicated in the field of computer science by the exact same scientific societies and journals. The Oxford Dictionary traces the history of the term in scientific journals published by a leading academic society in Electronics founded in 1884 around the development of electricity: IEEE, scientific society that gradually encompasses research in computer science.

- 1967 IEEE Trans Human Factors in Electronics viii 1/1 Multiple access on-line interactive man-computer systems.
- 1971 IEEE Trans Electronic Devices xviii 618/2 we can see that even from this simplified description of the composition of interactive terminals, a wide variety of disciplines are called upon to establish a successful interaction between man and the computer system.
- 1973 C.W. Gear Introd Computer Sciences iv 153 The input to the typewriter system is in the interactive mode.
- 1981 Event 9 Oct 28/4 Interactive video TV screens equipped with computer linked press-buttons for instant Q&A verdicts on the show – asked in the studio and answered in your own home.

The word eventually qualifies the exchange of information between a computer and its user: « pertaining to or being a computer or other electronic device that allows a two way flow of information between it and a user, responding immediately to the latter's input ». Examples then include the system user/computer:

- 1970 computers and Humanities v 24 It would be especially interesting to make this sort of investigation *interactively*, using some kind of terminal
- 1973 Physics Bulletin Aug 497/1 A computer program, Trace… can be interrogated *interactively* to show how store movements have progressed
- 1978 Scientific American Apr 26/1 Using satellite to transmit data, voice, full-motion and freeze-frame video, and facsimile documents – all *interactively* – Satellite Business Systems (SBS) has undertaken a pace-setting experiment in advanced communications for geographically dispersed organizations.

The rhetorical potential of the term, or to use the expression from the Cultural Dictionary of Science, its "suitably vague" scope aligns physical phenomena and human and social phenomena. It is interesting that computer science should be the place where this linguistic and conceptual crystallization happens. The Robert

[23] Encyclopaedia Universalis (2005).

Dictionary of French Language makes a hypothesis as to why such fields as electronic engineering and social science would merge. Because computers expand beyond the technical field of specialists and target new, larger, audiences, they achieve a social status and merge into people's lives: everybody can read, write, interpret and circulate digital texts. "Interactivity" has therefore shifted from mechanics, physical interactions, to man-machine interactions. While the expression is conveniently accommodating a diversity of connotations, it also supports a certain model of the system and how the system should work and be evaluated.

The criteria that are used to assess a computer system in the discipline are mostly gathered under the heading of "usability". The International Organization for Standardization *ISO 9126 (1991) Software Engineering Product Quality*, defines usability as: "A set of attributes that bear on the effort needed for use, and on the individual assessment of such use, by a stated or implied set of users".[24] *ISO 9241-11 (1998) Guidance on Usability*, defines usability as: "The extent to which a product can be used by specified users to achieve specified goals with effectiveness, efficiency and satisfaction in a specified context of use". Effectiveness means that the product allows the users to achieve their goal. Efficiency adds a notion of lesser effort or minimal time to achieve this result. All these criteria focus on the efficiency of the user within a system. The user is seen as a part of the system and has to perform just as efficiently as the artificial system does, without bugs, setbacks and disruptions. Eventually, some feelings are introduced in the assessment process since users should experience some satisfaction. The latter concept is quite vague. It relates to the comfort and subjective evaluation of the user. But the satisfaction of what and why is another question.

Introducing the word reflexive - as Pachet does - is therefore far from being a simple addition to a complex system. In subtle ways, it seems to me that it contradicts the interactive paradigm. I speak of paradigm because the use of the concept introduces not only to the properties of a specific machine but also to a whole system including humans and nonhumans, a model of activity, goals and values. Though these expectations moved forward the field of HCI and improved the performances of digital systems, we saw with the Continuator that certain systems can actually offer some space for a model of activity based on pleasure, reflection, and learning.

Even though the Continuator is not openly critical, it still makes a dent in the interactive paradigm and ideology.

[24] http://en.wikipedia.org/wiki/Usability

2.4 Resisting Interactivity

Some artifacts are openly more critical of the assumptions related to the reign of the interactive paradigm. The following art pieces do not play by the interactive rules and even seem to focus on contradicting them. They dispute expectations both about the technologies themselves and about our relations to them. These art pieces not only force us to reconsider our assumptions about technical systems, they also bring our own activities under scrutiny. In digital systems, the main assumption is that there should be a seamless and fluid continuity between the activity of the user and the reactions of the machine. The two artists in this section precisely explore what this means and how it could be different, and thereby open a new design space between humans and nonhumans. I picked up two French artists: Antoine Schmitt and Agnès de Cayeux whom I worked with or met several times and who have a very different style but a deep understanding of the digital media and the Internet.

2.4.1 Two Use Cases: "Just Married" and "With Determination"

Agnes de Cayeux's art work on eroticism is a case in point: her work deflects all assumptions about interactivity and at the same time plays with it. She shows that these technologies are primarily defined as efficient tools that act on the world. They are also artifacts that can affect us emotionally through a specific aesthetic experience. Finally they are media, opening a plane of representation that allows us to withdraw from actions and to reflect about them. This challenges the concept of interactive design as it points out that it is also representative and reflective design.

In *Just married,*[25] the network artist Agnes de Cayeux shows blurred erotic images that react only if the spectator slowly caresses the screen with the mouse or the touch pad. Attracted by the photos of naked bodies, willing to unveil more of the image or to capture the movements of the protagonists, the users want to hasten the whole process. But their expectations about the machine "real time" reactivity are thwarted by the maddeningly slow experiment. The image does change, revealing another scene but only after a long trial period. Agnes de Cayeux's work therefore confronts the spectators with their expectation of speed, feedback, effectiveness, efficiency and baffles expectations about real time interactivity. Time is made palpable again. Watching images does not necessarily equal consuming them as fast as one can click, but can be a gradual process of contemplation. Real time activity is no longer taken for granted, it is tried. Thereby, the spectator has to think about a number of things: the voyeurism directly related to expectations of reactivity from the machine, the primacy of efficiency over aesthetics, the mode of consumption of images and more generally of "cultural" artifacts.

[25] Cayeux (2003).

This work of art has explicit erotic references that not only stem from the choice of images and the use of a dictionary of erotic words but also because it reflects on the users' expectation of quick satisfaction that, for the artist, goes with interactivity as instant gratification. As Baudrillard analyzed in *On seduction*: "we are in a culture of early ejaculation".[26] In the '70s, the philosopher and sociologist already observed that our society of consumption encouraged expectations of easy and immediate reward. As couched in different terms by Virilio, it is a culture of speed and immediacy.[27] What the artist suggests is that, with digital systems, we are looking for a quick emotional rush that the computer is supposed to provide.

Agnes de Cayeux therefore obstructs the users' pursuit of efficiency, and rather invites them to focus on their own gestures, their manipulation of the system, and their relation to images. She tries to capture the way these technologies organize and shape the representation of the action as well as the action itself (Fig. 2.2).

A very different artist, Antoine Schmitt[28] has worked since the beginning of the 1990s on what he considers the "algorithmic material".[29] Antoine Schmitt creates animated figures endowed with behavioral algorithms. The forms and features are controlled by displacement and force algorithms and animated by random coefficients, producing unexpected changes. *With Determination* is a collection of 12 pieces: *Standing, Standing 2, Jumping, Not dying, Behaving, Not behaving, Resisting, Stepping, Stepping 2, Pushing, Not moving, Nailed.*[30] Notwithstanding their abstract nature, the programmed movements and their accidents encourage an anthropomorphic interpretation. The titles obviously pointed in this direction as well as the lines and ovals - strangely resembling disembodied limbs. As I asked students and spectators to use the website, they pointed out that these figures and their moves did express human "determination". They also expressed a feeling of connection and unease provoked by the movements of these forms that collide with the edges of the "frame" on the screen. Indeed, the figure collapses when it bumps into the frame, then picks itself up again (Fig. 2.3).

This schematic but anthropomorphic figure shows that realism is not necessary for pathos. By mimicking movements and gestures, schematic features are enough to show difficulty, alertness and fatigue. In effect, the question ultimately posed by "*With Determination*" is: are there mechanics behind humans[31]? Or, in the words of Antoine Schmitt, "what relations are there between the forces within us and those

[26] Baudrillard (1979).

[27] Virilio (1986).

[28] Graduated from the l'Ecole Nationale Supérieure des Télécommunications in Paris in 1984.

[29] From an interview with Sandra Vie, an MA student working on algorithmic art.

[30] Antoine Schmitt, "Avec détermination" (With Determination), September 2000 (source: www.gratin.org/as/avecdetermination/index.html).

[31] This reminds me of Bergson's book, *Laughter. An essay on the meaning of the comic*, The Macmillan Company, 1911: "The laughable element in both cases consists of a certain MECHANICAL INELASTICITY, just where one would expect to find the wide-awake adaptability and the living pliableness of a human being." p. 10.

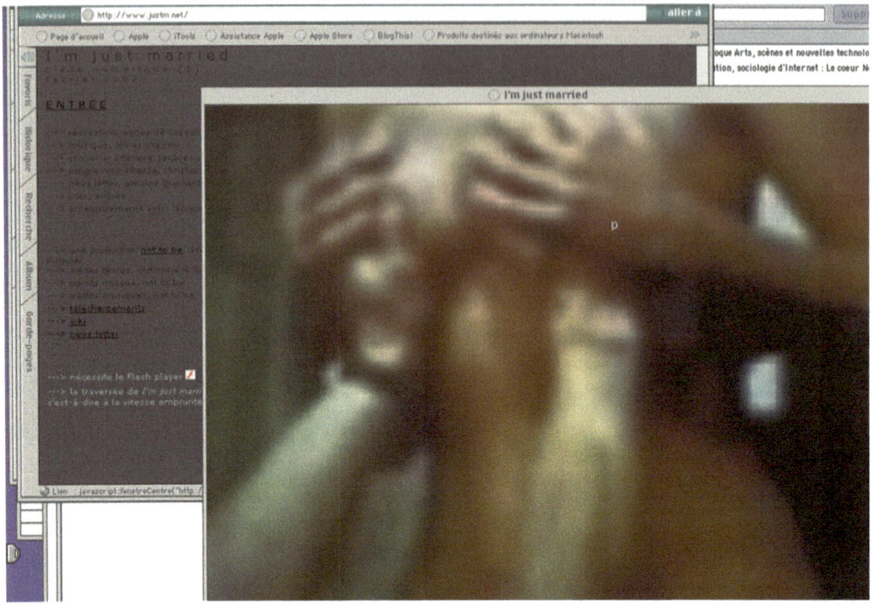

Fig. 2.2 Agnes de Cayeux, “I’m Just Married”, Connected Screens, 2003, (Source: http://agnes-decayeux.fr/AdC_site/AdCjustm/justm.htm)

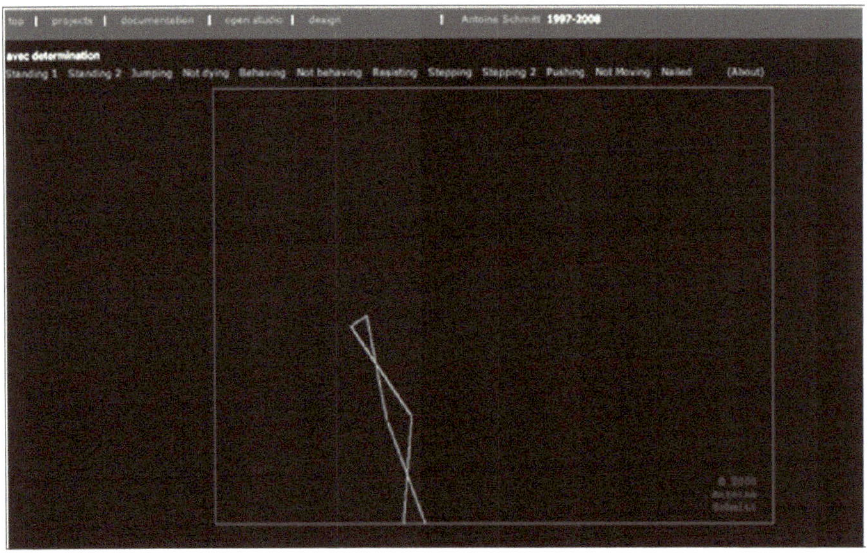

Fig. 2.3 Antoine Schmitt, “Avec détermination” (With Determination), September 2000 (Source: www.gratin.org/as/avecdetermination/index.html

outside of us? Where are we among these forces?" Schmitt explicitly raises the question of the aesthetics of moves and the relation between intentions and gestures:

> Why does it work like that? Perhaps this is the fundamental question. What are the forces behind movement? These movements have a form: which forces for which forms? Some forms have certain qualities (aesthetic, emotional, etc.). What are the qualities of underlying forces? What is their mode of existence? At what level do they act? What is the relationship between the forces within us and those external to us? Where are we positioned among these forces?[32]

But the visual is only one part of the work of art and could very well be compared to the famous experiment by Heider and Simmel who used an animated film with geometrical shapes moving around one another to demonstrate anthropomorphism.[33] In addition, Antoine Schmitt contradicts the users' actions on the system by having the figures respond in unexpected ways. In other words, he does not want to demonstrate some psychological traits but to highlight some cultural assumptions. As users manipulate the mouse, they provoke opposite effects from what they could expect. When they point upwards, the shafts fall, but when they point down, the shafts rise again. Looking at the screen is contradicted by manipulating it in the same way as the figures' movements on the screen are altered by invisible rules. If "*With Determination*" can create emotions such as compassion or embarrassment for the figures, users can also feel frustrated or annoyed at not being able to correctly manipulate the puppets. "*With Determination*" awakens the user's attention to gestures not only by representing them on screen, but also by associating them with unexpected events on the screen. In other words, spectators are given a distorted visual feedback and thus steered to look back at the way they move and actually question their moves. Assumptions about interactive systems are frustrated: the figures are autonomous and when they do react to users' input, they contradict the expectations of effectiveness.

In Agnes de Cayeux as well as in Antoine Schmitt's work, the interactive paradigm is both at work and challenged. All the associated qualities, like speed, efficiency, effectiveness, and satisfaction, are frustrated. As pointed out by Candy and Costello quoting Boehner et alii[34] who compare artistic practices and HCI practices:

> Artistic interaction design can be consciously non user-friendly, working to subvert expected norms so as to stimulate new interpretations, perceptions or behaviors (Boehner et al., 2005). The questions asked by these researchers are, therefore, often very different from those asked by interaction designers focused on usability in the work environment. Their different perspective is now proving to be very valuable as software systems increasingly expand out of the workplace and into all facets of everyday life[35].

[32] Schmitt, ibid.

[33] Heider and Simmel (1944).

[34] Boehner et al. (2005).

[35] Candy and Costello (2008).

One of the main consequences is that the user is made aware of her expectation about interactivity. The action/reaction system is not taken for granted, it becomes food for thought. This experience of interpretation where we observe and live the distance between our reflexes, assumptions, expectations and the actual output of the machine, is what I suggest we call a reflective experience. In the next section, I want to study the etymology of the word "reflective" in particular because it strikes me as being a phenomenological experience embedded in a particular aesthetic, that reflective designers are striving to achieve through their use of media.

2.4.2 *Defining Reflective in Design*

Reflective comes from the latin reflectere: "to bend back". The word reflective therefore comes from the practical spatial and temporal experience as well as encompasses the intellectual metaphor, as is the case in many Latin verbs that consider the process of thinking. For instance, a conversation is also cum-versare: that is returning back to a subject. Re-flectere is about thinking because it considers knowledge as an actual journey that explores an experience and comes back with a transformed concept. "animum reflexi" (Virgile En 2 741): "I have brought my thoughts back. I went back [to her] in thoughts, I thought [about her]". The Oxford English Dictionary mentions this second meaning: "Of mental operations: turned or directed back upon the mind itself," with the particularly apt quotation of 1640 by Reynolds: "In those two Offices of Reason, the Transient and Reflexive act, that whereby we looke Outward on others; or Inward on our selves."

For Barbara Babcock, who studied the notion in anthropology, religion, and literature, reflexivity is a specific intellectual practice that is related to any system of meaning making:

> The terms reflexive, reflexivity, and reflexiveness have been used in a variety of disciplines to describe the capacity of language and of thought—of any system of signification—to turn or bend back upon itself, to become an object to itself, and to refer to itself. Whether we are discussing things grammatical or cognitive, what is meant is a reflex action or process linking self and other, subject and object.[36]

While every living being makes sense of their experience, not every living being reflects. The difference lies in the capacity to use a mediated form of language that translates the activity as an object of enquiry. Humans can re-present the world and their relations and this very competence makes it possible to create not only a reflective space but a design space. Because our relation to the world is always a mediated one, through language and all the media that build and keep a distance precisely from "pure" embodied sense making, the interpretation and design space of signification turns the activity from a problem of ergonomics to a question of semiotics.

---

[36] Babcock (1980).

Strikingly enough, the metaphor of reflectivity brings together the physical with the psychological just as the metaphor of interactivity. Indeed, it brings together several meanings:

- The actual function or movement that goes back to the point of origin as a reflex is a bodily answer to a stimulus. The real time response in digital system is reflex-like.
- The propriety of a surface. A surface is reflective if it can send back something like an image in a mirror. The screen that represents the response of the system to our input is this reflective surface.
- The capacity for contemplation and deep thinking. Media exist as a plane of representation that not only offers information and structures contents, but in the case of information technologies, can question the user's assumptions about the system.

Designing digital systems is an interesting challenge because the designer not only questions assumptions as she would with any other artifact, but does it with a device that can actually share this reflection with the user because it mediates the activity. The screen as a space of representation is also the space of action. Contradict one or the other and the user is immediately pushed into reflecting on what the system is not only doing but telling her. A reflective digital device brings together designer and user to reconsider their values and assumptions because it stages these values. While it may seem that the designer "just" adds these different points of view, in the next section I want to develop how the reflective and the interactive dimensions are more than surface properties of a technical system but support at least two models of human activity. Very different worldviews are brought together that we need to compare.

2.4.3 Comparison Between Interactive/Reflective Metaphors

From the standpoint of interactivity, digital media function as reflex technologies. They are seen as an extension of the body, tools that increase the body's efficiency to act on the world, and that amplify human actions. In reflex technologies, time is a fundamental parameter. Technologies that work as prostheses tend to disappear behind the action so that there is no delay between what we do and the result of our action. Reflex technologies are meant to be transparent to action. They support human activity because they have a model of the user, of the task, and of the system that is sufficiently seamless for the activity. The paradigm of interactivity also supports a model of design that is problem solving. In this perspective systems have to solve problems and optimize the means to support the user's ends. But what if the model of activity is wrong or what if the values do not suit the users? What if the system is yet another way to aggravate gender issues, or power discrepancies? How can designers actually extract themselves from the ideology that naturalizes "efficiency", "transparency", and "real time"?

These considerations have increasingly been incorporated by the developers' community as they try to assess the impact of devices on the user. The next step was to actually consider that efficiency was to be paired with pleasure. Gradually, HCI was concerned with the experience of the user and in particular her emotions.[37] Ergonomics now deal not only with the direct physical impact of technologies on the user's body and more and more with the psychological cost of using a technology. From a design perspective, objects can seduce users.[38] Human beings look beyond functionality to aesthetic pleasure. John Heskett, historian of design, structures his introduction to design along this tension between functionality and pleasure and the way they can converge:

A primary condition of utilitarian design is that it must effectively execute or support certain tasks. In contrast, a piece of jewelry, a porcelain figurine, or a frame for a family photograph has no such specific purpose – instead their purpose can be described in terms of contemplative pleasure or adornment.[39]

But this does not promote the reflexivity that is needed to study, criticize, and more importantly move beyond social habits, technologies ready at hand, and worldviews more generally. More to the point, by making artifacts or interfaces more pleasurable, designers in effect try to make sure that nobody complains about or contests these technologies.

What makes media so special compared to other artifacts is that they support our capacity to distance ourselves from action and reflect on our experience. The epistemological cut between "being in the world" and "representing the world" is crucial to the understanding of what these technologies mean for human beings but also for designers. As design tools, they open a design space which is the foundation of our reflection. Reflective technologies therefore not only are based on a model of teleological action and of psychological and cultural aesthetic experience, they also open a space of representation. Using the reflective metaphor not only helps interaction designers consider information technology as new media, a specific milieu for reflection, it also helps them consider more broadly what is needed in the design process: a place, or means, to distance themselves, or even extract themselves from the obviousness of everyday use.

2.4.4 Reflective Technologies as a Metaphor for the Design Process

To elaborate from the interactive metaphor to get to the reflective metaphor means that we change the agenda from engineering to political. Like books, TV, or the cinema, information technologies are instruments for debate. Different

[37] Grudin (2012).

[38] Bannon (2011).

[39] Heskett (2005).

infrastructures and systems of information allow different people in different circumstances to participate in this debate.[40] Newspaper or flyers, tweets or Facebook pages, all play a part in the way we organize political participation as we will see more in depth in Chap. 5. Reflective design is not only about people participating, it is also an aesthetic exploration. It means that we allow different standpoints and formats. Media are therefore a central concept to the practice of design as specific tools for the complex denaturalization process of cultural forms and activities. Designers use these technologies not only because they act on the world or because they affect us, but because they specifically open a plane of representation and a design space.

The following table summarizes the interactive and reflective paradigms. The first column gives five points of entry in the reflective paradigm. First, as artificial constructs, information technologies are part of our relation to the world. The first line addresses the way we relate to the world. The second line introduces the model of action that is the basis for the development of these technologies. The third line presents some key words that describe design goals. The fourth line addresses the qualification of information technologies (IT) and in particular the underlying epistemology that defines them. The fifth line gives a succinct viewpoint of the disciplines that are primarily concerned. Though a simplification of what different

	Interactive paradigm		Reflexive paradigm
1- Main relation to the world	Action on the world	Passion (how we are affected by the world)	Representation, mediation of the world
2- Rationale of our activity	Teleological rationale	subjective and social rationale	communication rationale
3- Model of the technology	tool and prosthesis	artifact	text, document, media
4- Design focus	Activity	Taste - Emotion	Interpretation
5- Key words for the design task	Problem solving, optimization	Ergonomics, seduction	Literacy
6- Figures of the « user »	Multi-tasker	Aesthete	Homo sapiens
7- Design evaluation	Effectiveness, Efficiency, learnability,...	Cultural adequacy Social adequacy	Readability, communicability

Fig. 2.4 Overview interactive/reflective paradigms

[40] Rheingold (2000).

disciplines bring into analysis and design, the framework highlights methodologies and concepts that are involved (Fig. 2.4).

2.5 Designing for Reflection: Two Use Cases in Distributed Networks

While the first three examples described completed artifacts and their critical standpoints, the following two examples focus on the design process and show how the plane of representation was crucial.

I chose these two research projects because they were the results of "a reflective design by accident". Rather than hiding the workings of the system, the first experiments actually showed the users what was happening in their devices. These accidents actually taught us a good lesson: to be able to question the value of a system beyond mere functionality, it helped that users had insight into how it worked through the display of information on the screen.

2.5.1 Quick Overview of the Projects

Both research projects that are presented in this section pursued approximately the same goals. Between 2003 and 2005 the research project Safari[41] was looking into developing Mobile ad-hoc networks (MANET) that are self-configuring networks of mobile nodes connected by wireless links (such networks may be obtained by using Wi-Fi cards in ad-hoc mode) as well as possible applications. One of the partners was the SNCF, the French National Railway System and the MANET experimentations were to be deployed in one of the main railway stations in Paris (Fig. 2.5).

> SAFARI aims to create, combine and perform protocol infrastructure and software necessary to access transparent, automatic configuration, integration and adaptation services over an IPv6 network in ad hoc mode with the wired access. Its contribution is the design of innovative software solutions and protocol based on existing and proven standards (IPv6, multicast, proxy applications, programmable networks, …) capable of responding to dynamic constraints on both the network infrastructure and the supply and continuity of services.

The project's main economical concern was to develop access to mobile services. It was targeting a way to embed technologies that would no longer depend on fixed and costly infrastructures.

[41] 2003–2005, ANR –RNRT, adhoc wifi research project SAFARI. http://codesignlab.wp.institut-telecom.fr/research-projects/safari/

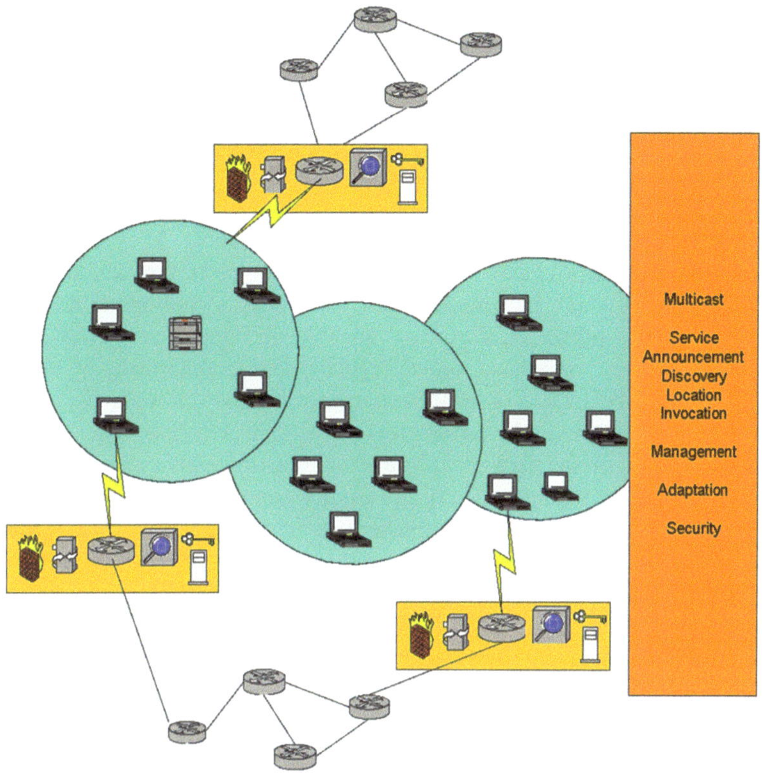

Fig. 2.5 Architecture of the Safari project

The project "Transhumance",[42] between 2006–2008 focused on developing a specific middleware for MANET.[43] Major issues were connection, reconnection, and disconnection of mobile nodes and the fact that routing information rapidly drained the battery of the mobile devices.

> Mobile ad-hoc networks bring new constraints: in particular, since nodes act both as end-user terminals and as routers and since they are mobile, they may become out of reach (for a short or long period of time, or even permanently). This must be addressed by the middleware in order to ease the development and the deployment of applications. The middleware must also provide applications with common features such as communication mechanisms, resources discovery and security management. Energy management is also a crucial issue for battery operated mobile devices[44].

Security and energy management challenges were often identified as key issues, but were never completely addressed in a fully integrated and operational solution.

---

[42] 2006-2008, ANR –RNRT, adhoc wifi research project Transhumance. http://codesignlab.wp.institut-telecom.fr/research-projects/transhumance/

[43] Gentes et al. (2008).

[44] Gentes et al. (2008).

To test middleware and software, as well as contemplate the social stakes of the infrastructure, two games were designed. The first one, Treenor,[45] based on the visualization of a spontaneous network on a pocket PC (iPAQ WiFi H5550), was a "shooting" game taking place in Gare Montparnasse (a railway station in Paris) where actual train arrivals and departures served as the starting point for a virtual battle of space crafts (tests were conducted in January and February 2006, with 20 testers. There were observations and in-depth interviews with the players). In Treenor, players chose a clan of either Humans or Robots. They could visit the Treenor planet that reproduced the architectural features of the station. They met other players with whom they could chat if they were within adhoc reach. When players wished to embark on the active phase of the game, they consulted train timetables (that were, otherwise, shown regularly and automatically on their interface) and went to the dock to board the train/spaceship they chose to "fly" with. The game was synchronized on the actual schedule of train arrivals and departures. The player had to stand on the platform of the actual train to be able to board the virtual spacecraft. Then, the second phase of the game started where Humans and Robots chased each other. In this phase, the player was in control of his armed spaceship and faced the opposing clan. Players of the same team had to fight and kill the enemy's spacecrafts. If neither clan was wiped out before the actual departure of the train from the station, the game ended for all players positioned on the platform. They landed back on Treenor to take a new train/space craft. A striking feature was the synchronization of virtual and real world along with actual events happening in the railway station, like the arrivals and departures of trains. The blending of reality and fiction was unanimously noted by the testers to be the highlight of the game (Fig. 2.6).[46]

Team Exploration,[47] also based on a MANET[48] but on a PDA (Nokia N770), offered to canvas a scenic neighborhood of Paris (La Butte aux Cailles). People had to gather the pieces of a puzzle made of architectural and urban details (Tests conducted in 2008 included 16 players, observations and in-depth interviews). The Transhumance project[49] tried to address a certain number of issues that were pointed out by the players and testers of Treenor, more specifically on how to provide rich contents on the premises. The game consisted of finding a mysterious meeting place thanks to clues scattered in the historical Parisian area of "la Butte-aux-Cailles". Two teams of four players each were opponents and had to gather as many clues as possible and be the first to reach the mysterious place to win the game. The interface allowed team members to communicate among themselves and to collect the answers. The whole game was designed to let the players be disconnected from their

[45] 2003–2005, ANR –RNRT, adhoc wifi research project SAFARI. http://codesignlab.wp.institut-telecom.fr/research-projects/safari/

[46] Gentes et al. (2009).

[47] 2006–2008, ANR –RNRT, adhoc wifi research project Transhumance. http://codesignlab.wp.institut-telecom.fr/research-projects/transhumance/

[48] Mobile Adhoc NETwork.

[49] 2006–2008, ANR –RNRT, adhoc wifi research project Transhumance.

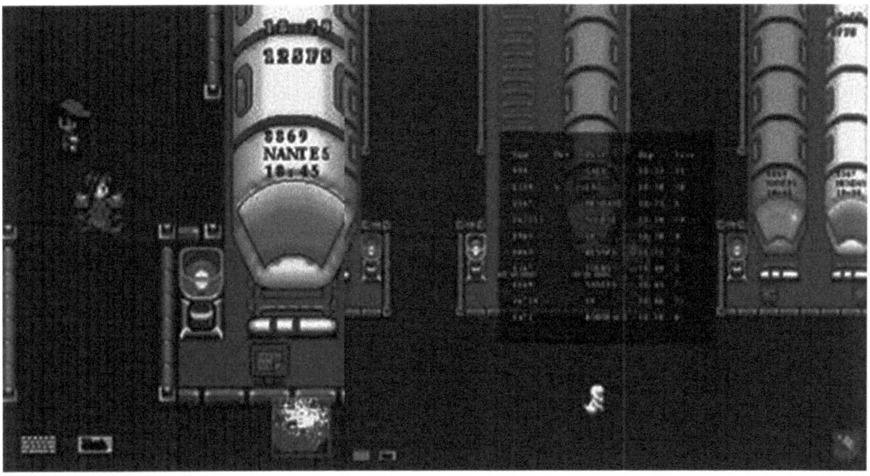

Fig. 2.6 Screenshot of Treenor Game ©2003–2005, ANR –RNRT, SAFARI

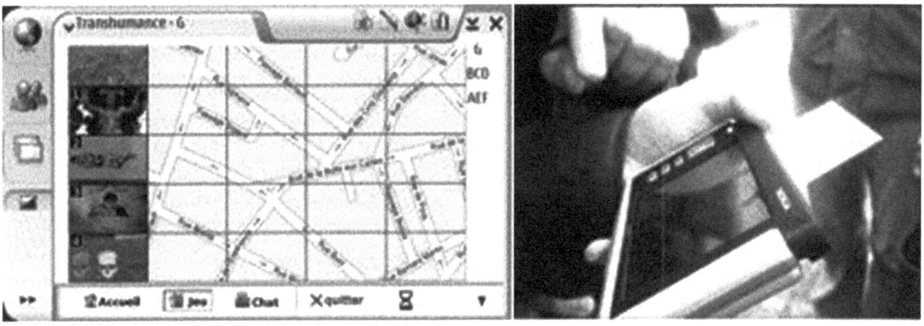

Fig. 2.7 Screenshot of Team Exploration Game – Players in action ©2006–2008, ANR –RNRT, Transhumance

team so that they could freely explore new premises. When a player came close to another one a connection was immediately established. During the game, players lost and got the connection back, although, to validate an image, all members of the team had to be connected. The interface of the game was provided by a map of the Butte-aux-Cailles partitioned in twenty rectangles. On the left of this map, five pictures were displayed. The top one appeared "blurred", it was the final meeting place (Fig. 2.7).

In "Team Exploration", as in "Treenor", there was no geolocalization. The feeling of getting to know the city was based on the discovery of fine details (different from a tourist's approach to major buildings) located on a map. There was a limited time to locate the pictures and when a suggestion was made, it had to be approved by the other members of the team through the game interface.

2.5.2 The Bug That Saved Us

The engineering research goal was to have a network without any of the telecommunication infrastructure: no cables, antennas, nor satellites. In a MANET, the key expression from an engineering perspective is that **the devices are the network.** The questions are: how does the network reconfigure itself while nodes appear and disappear because people move with their devices and can be out of reach and how do systems avoid losing data in the process? At first sight, these questions presented nothing that would be of interest for the end user, but a bug in the first project Safari helped us go beyond this perspective and not only did we design our new application in Transhumance very differently, but we also came up with new questions.

While playing "Treenor", some testers lost the end-user interface and suddenly got the OS interface that displayed the status of the network. In particular, they could see the routing with an interface representing "hops" between players. They were intrigued and as the research team helped with the bug, they asked to play with it so as to see how the configuration changed while they were moving within or outside the range of another device. A pretty exciting moment occurred when the testers realized that they had access to a more distant PDA through another closer one. The three testers who had the bug then looked around to try to identify carriers of the other PDAs. They did not find them, but curiosity and impatience to find the other group was explicit. One of the testers asked if we could give names to identify the PDA by something other than its IPv6 address. Suddenly, another important asset of the game was that the technical network and people were fused.

From our testers' standpoint therefore, "**people were the network**". In Transhumance, we then decided to show a topology of the network.[50] Displayed on the right of the screen in Team Exploration, it showed the connections to other players (and the number of hops to a player). This change in the interface had at least two major repercussions. As we describe below, first, participants suddenly considered their physical position as part of a technical process; second, they realized that they had to share their battery by forwarding messages.

The game changed how people considered their responsibility in relation to the technology. When describing their experience, the testers used three metaphors: the human infrastructure "works as a bridge", "a chain", a "relay". Testers said: "the person is in the center. The adhoc techno breaks down barriers and builds bridges". R, one of the testers, was disappointed that: "Team exploration has not exploited the multi-hop" even though he believed that it had "great potential." He saw a "strategic interest" by creating a "chain of the same team" that would "get information along the chain of his team". Testers said they considered themselves to "be a relay". Another tester thought of fun activities that could exploit this string of nodes: "the Chineese Whispers/telephone" that relies on the idea that each link has a role in the passage of information. All these metaphors clearly merged people and technology. In their experience, people "were" the infrastructure.

[50] Demeure et al. (2008).

But, one of the main issues is that every routing of messages depletes the battery of the device. One tester contended that she would "spend to give back to my network", but others clearly hesitated in being generous to the point of being incapacitated from using their own device. The problem was also the bandwidth. In Safari, the tests showed that downloads were extremely slow or impossible when the bandwidth was shared. The experience of having to share bandwidth was extremely unpleasant for the testers. Each user wished to be in "priority". In Transhumance, a lot of thought was given to the subject of energy consumption on a technical basis[51] but also, thanks to Safari, on a design basis. This project had allowed us to observe the not so positive reactions to the loss of energy from users who were appalled by the short life span of their devices in ad hoc mode. "Being the network" through physical positions therefore affected the testers in both positive and negative ways. The extra bit of attention needed to rebuild connections between participants and the fading battery were either considered a fair price to pay for the instantaneous constitution of a network or a burden a little too heavy for the benefits. Before the final version of Team Exploration, we even thought of giving the players a full view of their energy consumption, showing them how different types of activity would more or less deplete their battery. The whole game could have revolved on strategic decisions whether or not to communicate for instance, or whether or not to allow others to use one's device as a node. We eventually decided against it, as we also wanted to encourage users to enjoy relating to each other and not necessarily resent it. In the design of the game, we ended up choosing the time allotted to the party, so that the devices would not run out of energy before the end of the game. It was a way to avoid the problem but it also stopped the design exploration that tests of this scenario would have given us an opportunity to pursue.

The following table sums up the different viewpoints and design stakes of these two projects (Fig. 2.8).

The shift from an "engineering" viewpoint that considered "devices as the network", to a "user" viewpoint that considered "users as the network", finally led to the interface designer's question: **how do we represent people/the network?** How are users represented so that the interface still provides private, secure, intimate, as well as public communication? How does it also provide an understanding of space and location while using the new concept of "relative distance"? These questions - that were fundamental to the innovation - could not be asked without building an interface that showed some of the options. Representing people and a form of topology of the network was necessary, not only to enable strategic moves, but so that testers could actually reflect on the values of such a system. The bug saved us from not having this discussion.

---

[51] "The *energy management* block involves a monitoring module and a decision module. The decision module decides, based on a policy and on the information about the energy collected by the monitor, of the adaptation actions to be executed. The possible actions are implemented in middleware modules and consist of adaptations of their behavior that reduces the energy consumptions (e.g. stopping messages acknowledgements in the transport protocol)".

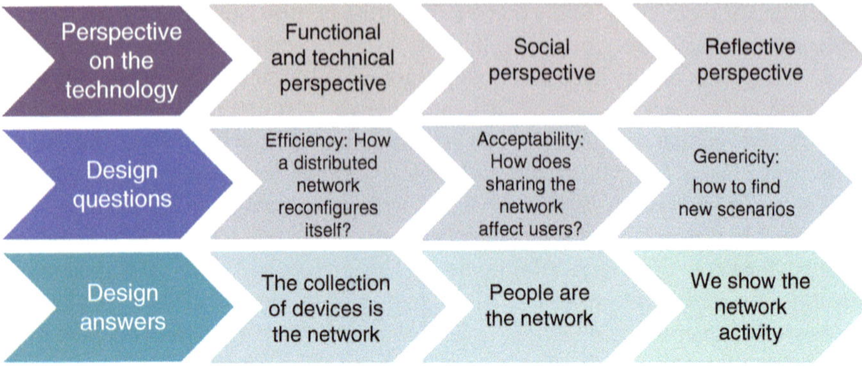

Fig. 2.8 Design perspectives of Safari and Transhumance projects

In both projects, the final interface was not the design goal. People could easily argue that these games were not so innovative, that the graphic design was lousy, or that the game play was minimal. But this is beside the point. The interface served as a design space where new questions could be asked. It expanded the technical, social, and aesthetic perspectives.

2.6 Reflective Objects for Reflective Design: A Delicate Balance of Signs

From the last two examples, one could deduce that dysfunction triggers invention and some sociologists and philosophers develop this hypothesis. But from a design perspective, it gives too much space to chance and one can legitimately wonder how to deliver more reflexivity without failing so much.

2.6.1 Reflectivity as Method

Sociologists in particular sociologists of infrastructures[52,53] and researchers in design[54,55] repeatedly point out that only a dysfunctioning system forces people to step back and reconsider all the elements of a situation: the technical aspects of the system, the designers' intentions,[56] and their own expectations. The definition of

[52] Star (1999).

[53] Star and Ruhleder (2001, 305).

[54] Suchman (1987).

[55] Norman (2002).

[56] Crilly (2011).

reflectivity in these instances is based on phenomenology. People consider what is used, as opposed to what is looked at, based on the "ready to hand" / "present at hand" opposition conceptualized in Heidegger's study of our relation to the world in Time and Being.[57] This conception of reflectivity relies on an understanding of human activity as goal oriented and things as tools[58] whatever the shape, material, consistency of the artifact. From this point of view, reflectivity is based on the experience of the user and more precisely by what Heiddeger calls the "unreadiness to hand". For different reasons the user's activity is challenged by an environment that does not respond as it should. But reflecting about what is going wrong is not similar to expanding the meaning of a device or a system. Dysfunctioning as a way to step back and look more closely at systems can be the first step to conceiving a new one but the phenomenological explanation must be completed by a semiotical explanation for two reasons: first, to offer alternative scenarios is not to dysfunction, second, alternative scenarios and dysfunctioning scenarios are in my opinion based on a transformation of our relation to artifacts from a practical to a semiotical one. Indeed, in design the dysfunction is not only frustrating, it is the starting point for a reorganization of material that depends on this distance that turns the system into a semiotic plane of interpretation. The technical elements become signs that can be read and recomposed and not simply used. In other words, the input/output of the system becomes what is represented. The first examples of this chapter – "The Continuator", "Just Married" and "With Determination" - are not dysfuntionnig. Nonetheless, they transform functions into signs with a meaning. For instance, Agnes de Cayeux no longer uses "real time" speed but then time becomes a subject of enquiry. That is what happened with our application "Treenor" within the Safari project where positions in space also became a new question. Finally, I think that the Continuator also strikes a fine balance between playing and listening, acting and reflecting. It keeps a tension between the activity and foregrounding its modus operandi. However, from these different examples, one can conclude that some artifacts are more reflexive than others. This leads me to my second concluding remark that I develop in the next section: reflectivity is not only a method it is also a question of style.

2.6.2 *Reflectivity as Style*

The same media can be designed to provide a total immersion in the narrative or the action. People in the rush of action, playing intuitively with their image, or adjusting finely to social norms think about what they do but do not necessarily engage in a reflection about what they do. The pilot in a fighter wants to be harnessed to a technology and visualize information that is for immediate use not for philosophical introspection. While everybody can reflect on it, the object itself does not give a clue that there is space for improvement, or for an altogether new paradigm. The point

[57] Heidegger (1962).

[58] Harrison (1996).

that I want to make here is that certain designs will foster more reflectivity than others. They will stage more of a meta-discourse than others. Authors might use language and media in such a way that it is not transparent to reading nor to acting. Thus, two questions arise:

- How will interactive systems translate the user's activities into a representation of her inputs so that she can reflect on them?
- How will interactive systems support the balance between immersive activity and reflection?

Thinking of the design process, I want to argue that designers organize this reflexive space to break away from traditional values and artifacts. In this section, as we stir towards style and forms, I find the discussion on reflexivity in the humanities to be particularly useful because it tackles the aesthetic issue of stylistic invention.

Art historians and theoreticians of literature note how art and narratives can be absorbing for the reader or spectator.[59] The whole system tries to organize "this suspension of disbelief for the moment" (Coleridge[60]), the perfect viewpoint (perspective that will give the spectator the illusion of being inside the painting or the story[61]). Display devices always work on refined immersion systems like the darkness of theater and cinema that directs the attention of the viewer. But, certain writing and visual practices also tend to promote the workings of the media over the illusion of the story. Since the romantic era, it has helped to consider that art is not a reproduction of reality but reflects on itself. Oliver Grau who studies the concept of immersion in art notes that: "since the eighteenth century, aesthetic theories have regarded distance as a constitutive element of reflection, self-discovery, and the experience of art and nature[62]". The very much debated art for art's sake reminds the spectator or reader of the textuality of the contents and status of the work of art as a media construct. Reflexivity then means the process by which texts (or movies or any art form) **foreground** their authorship and production: "Reflexivity subverts the assumption that art can be a transparent medium of communication, a window on the world…[63]". In effect, there are different strategies of interruption and discontinuity that draw the reader or spectator's awareness to the fact that he or she is confronted with an intellectual construction. From stylistic virtuosity, narrative discontinuity, authorial intrusion, essayistic digression, to staging the production process (some equipment that gets in the camera field of vision) reflexive emphasis draws the spectator's or reader's attention to the formal materials and processes of media construction, by displaying the tricks of the trade, or introducing all kinds of violation with the narrative codes or genres.[64] Reflectivity therefore presents stylistic traits that come from but stand out from common language. In this respect,

[59] Arasse (1997).

[60] Coleridge (1985).

[61] Iversen (2005).

[62] Grau (2004).

[63] Stam (1992).

[64] Pearson and Simpson (2001).

reflectivity is grounded in semiotics. It is not only a perturbation of the activity. It is the variations on styles that attract the attention.

Distinguishing and differentiating semiotic traits is a necessary step to consider the otherwise banal artifacts as supply pools of aesthetic, social, and technical properties that can later be reused, re-composed, for a new design. In the literature on style and genres, attention is paid to how to define that which stands out of the ordinary. From the very beginning of poetry and rhetoric, there is an understanding that everyday language serves the purpose of social activities. What stands out is, therefore, any invention that draws the attention not only to what is said but how it is said. In the rhetoric of Quintilian,[65] these forms depart from the "usual" conventions. The literary theoretician Tzvetan Todorov considers that to study genres is to study the discourses about genres: "I believe we will have a useful and operative notion that remains in keeping with the prevailing usage of the word if we agree to call genres only the classes of texts that have been historically perceived as such[66]". This theory pragmatically anchors genres in society. This does not mean that there is no way to define what a specific genre is, but rather that we must be interested in the rational arguments that people provide to describe different genres. Most of these arguments are based on a description of the formal characteristic of texts and can be related to an underlying theory of emotions and human expression. Todorov points out that from Aristotle's poetics to Diomede, three main genres are presented as « natural »: lyrical, dramatic, epic. But Gérard Genette shows how these categories are extremely complicated to apply to a considerable number of literary works, "perpetuating unease and confusion.[67]" On the other hand, he emphasizes the fact that a theory of genres opens up new ways of interpreting different texts and even of perceiving some original text that would have escaped notice otherwise. Genette uses the word architext to describe a structure that supports reading and interpretation. In fact, genres tend to be structured for a reader. That is what Schaeffer calls "a retrospective projection.[68]" Actors of the art world pick up traits that they put together to constitute: « an ideal text of which all the real empirical texts are more or less faraway echoes[69]». Styles and genres are, therefore, defined after the fact by critics and readers who gather different texts that they find similar in certain ways. They can also be explicitly used by the writer or artist who knows that they can play on the social expectations of readers/spectators. Speaking of genres is a way to constitute a field, and its boundaries and therefore to signal how to trespass. The reader and spectator mobilize their knowledge of codes and literacies to understand the work of art, only to be destabilized in their expectations by the defaults of this archetype.

[65] Quintilian (2002).

[66] Todorov (1990).

[67] Genette (1979).

[68] Schaeffer (1989).

[69] Ibid.

2.7 Conclusion: Reflective Design with Reflective Artifacts

Reflective design is about methods and objects: the reflective practitioner needs reflective artifacts to question her design moves. First, as stated by Kroes,[70] designers need to be aware of the formal qualities and properties of the artifacts that they create. Second, interactive or reflective technologies could very well be understood as a metaphor of the design space in general. Their way of interacting with us in this tension between action and representation introduces the necessary space to reflect on what we are doing. It is our hypothesis that, as such, these technologies can show us what the design process is. Third, they can be used as simulation tools for that very reason. Using different software to accomplish the same task, or any other form of comparative performance exercise show alternative results, deviations, and possible fruitful differences.

There is, therefore, a double need to understand what these reflective artifacts are, not only because certain artifacts are "by nature" reflective, but because design practice is about stepping back to understand what one is doing, what are the options, what are the embedded values. The designer can face and manage alternatives between advancing solutions and questioning them not only when she works with the different media of design (like sketching or narrating), but also when she considers artifacts and situations as a semiotic plane of composition and interpretation. Such a space is a safe way to play with different options. Artifacts at that stage can be designed so that they are not yet totally streamlined. They are, so to speak, in defiance of "easy" answers.

In the examples taken so far, The Continuator, Agnes de Cayeux and Antoine Schimtt works of art, and the interfaces for distributed networks, a balance is finely struck between "simplicity and control". Simplicity in the sense that users manipulate the musical instrument, or the image on screen without having to actually learn a special interface. Control in the sense that their "spontaneous" gestures are not seamless: users get unusual feedback on their actions that lead them to think about them. These examples show the inherent tension between flow[71] and immersion versus distance and reflection that pervade all digital interactive systems. They also point to the fact that designers have a choice to foreground or not the reflective part of the system, to produce seamful or seamless interfaces. This appears to be a recent concern for the wider HCI community as well.[72]

Researchers in HCI now consider that "designing for reflection is becoming of increasing interest[73]". They contend that a number of computer systems need to support self-reflection in their users, in particular when health issues are at stake that need a change of behavior. In 2009, the CHI conference held a special workshop on technologies that supported reflection in experience: "to explore the movement

---

[70] Kroes (2002).

[71] Csikszentmihalyi (1991).

[72] Sengers and Gaver (2006).

[73] Fleck and Fitzpatrick (2010).

from designing for experience as interaction with technology, towards designing for reflection on felt-life experience captured by technology". Reflective design is showing something that then can be discussed. In the distributed network experiences, we could have missed a very important discussion by not showing the inner workings of the technology. As pointed out by Hans Gelter reflective thinking is "uncommon[74]" but it is critical to the design process when one needs to come up with new artifacts or services.

In design, reflexivity is therefore the concept that focuses on the way designers play with the different levels of meaning through objects specifically designed to further new questions about functionalities, personal emotions, social rituals, and knowledge. In the rest of the book, I will develop the different aspects of reflective design as a generative activity that strives to bring alternative and critical perspective into the creative process. For now, I want to emphasize what this concept means in terms of a project regarding time and actors. Reflexivity means taking the time to look back at different elements before they are totally integrated. Serendipity in the design spaces, tools, or objects is reflected upon before being put to use. The designer's stage is therefore not a power tool shed but rather a consciously organized environment that will make this expansive reflectivity possible. This also brings a major change in the way that the first users of these technologies can participate in the design process. While they can be considered as part of an action based system, they can also be considered as part of the interpretation/invention process. The latter option means that design does not only attend to their task and pleasure but gives them food for thought. A new layer of interaction with this "homo sapiens" is therefore necessary as the next chapter addresses.

References

Arasse, D. (1997). *Le détail. Pour une histoire rapprochée de la peinture*. Paris: Flammarion.

Babcock, B. A. (1980). Reflexivity: Definitions and discriminations. *Semiotica, 30*(1–2), 1–14.

Bannon, L. (2011). Reimagining HCI: Toward a more human-centered perspective. *Interactions, 18*(4), 50–57.

Baudrillard, J. (1979). *De la séduction*. Paris: Galilée.

Boehner, K., DePaula, R., Dourish, P., & Sengers, P. (2005). *Affect: From information to interaction, critical computing* (p. 59e68). Arhus: ACM.

Bosma, J. (2011). *Nettitudes: Let's talk net art*. Rotterdam: NAi Publishers.

Candy, L., & Costello, B. (2008). Interaction design and creative practice. *Design Studies, 29*(6), 521–524. https://doi.org/10.1016/j.destud.2008.07.002.

Cayeux, A. d. (2003). *I'm just married*. Paris.

Chartrand, T. L., & Bargh, J. A. (1999). The chameleon effect: The perception-behavior link and social interaction. *Journal of Personality and Social Psychology, 76*, 893–910.

Coleridge, S. T. (1985). *Biographia literaria: The collected works of Samuel Taylor Coleridge, biographical sketches of my literary life & opinions* (First Paperback Edition). Princeton: Princeton University Press).

[74] Gelter (2003).

Cooper, A. (2004). *The inmates are running the asylum: Why high tech products drive us crazy and how to restore the sanity* (1st ed.p. 29). Indianapolis: Sams – Pearson Education.
Crilly, N. (2011). The design stance in user-system interaction. *Design Issues, 27*(4), 16–29.
Csikszentmihalyi, M. (1991). *Flow: The psychology of optimal experience* (1st ed.). New York: Harper Perennial.
Demeure, I., Gentes, A., & Guyot-Mbodji, A. (2008, May 19–23). Transhumance: A platform on a mobile ad-hoc network challenging collaborative gaming. In: *Proceedings of the international symposium on collaborative technologies and systems*, Irvine, California. New Jersey: IEEE Press.
Encyclopaedia Universalis. (2005). *Dictionnaire des Notions*. Paris: Encyclopaedia Universalis.
Fleck, R., & Fitzpatrick, G. (2010). Reflecting on reflection: Framing a design landscape. In *Proceedings of the 22nd conference of the computer-human interaction special interest group of Australia on computer-human interaction*, 216–223, http://dl.acm.org/citation.cfm?id=1952269
Gelter, H. (2003). Why is reflective thinking uncommon, *Reflective Practice, 4*(3), 337–344.
Genette, G. (1979). *Introduction à l'architexte*, SEUIL (Seuil).
Gentes, A., & Mollon, M. (2015). Critical Design: A delicate balance between the thrill of the uncanny and the interrogation of the unknown. In *Empowering users through design: Interdisciplinary studies and combined approaches for technological products and services* (New York: David Bihanic, s. d, pp. 79–101). Cham: Springer International Publishing.
Gentès, A., Guyot-Mbodji, A., & Demeure, I. (2008). Gaming on the move: Urban experience as a new paradigm for mobile pervasive game design. In *Proceedings of Mindtreck 2008, ACM, Tampere, Finlande Mindtrek 2008*, Tampere, Finlande, 6–9 Octobre 2008.
Gentes, A., Jutant, C., Guyot-Mbodji, A., & Simatic, M. (2009, June). RFID technology: Fostering human interactions. In K. Blashki (Ed.), *Proceedings of IADIS international conference game and entertainment technologies 2009* (pp. 67–74). IADIS Press.
Grau, O. (2004). *Virtual art: From illusion to immersion*. Cambridge, MA: The MIT Press.
Greene, R. (2004). *Internet art*. London: Thames & Hudson.
Grudin, J. (2012). Introduction: A moving target: The evolution of HCI. In J. A. Jacko (Ed.), *Human-computer interaction handbook: Fundamentals, evolving technologies, and emerging applications* (3rd ed.). New York: CRC Press.
Harrison, S. (1996). Re-Pl ace-ing space: The roles of place collaborative systems. *Citeseer, 7*, 67–76.
Heider, F., & Simmel, M. (1944). An experimental study of apparent behavior. *The American Journal of Psychology, 57*(2), 243–259. (avril 1944).
Heidegger, M. (1962). *Being and time* (trans: Macquarrie, J., & Robinson, E.). New York: Harper and Row Publishers.
Heskett, J. (2005). *Design: A very short introduction*. Oxford: Oxford University Press.
Iversen, M. (2005). The discourse of perspective in the 20th century: Panofsky, Damisch, Lacan. *Oxford Art Journal, 28*(2), 191–202. Oxford University Press.
Jeanneret, Y., & Souchier, E. (1999). Pour une poétique de l'écrit d'écran. *Xoana, 6–7*, 97–107.
Johnston, A., Amitani, S., & Edmonds, E. (2005). Amplifying reflective thinking in musical performance. In *Proceedings of the 5th conference on creativity & cognition*, pp. 166–75. http://dl.acm.org/citation.cfm?id=1056248
Kroes, P. (2002). Design methodology and the nature of technical artefacts. *Design Studies, 23*(3), 287–302.
Le Marec, J. (2001). « Dialogue interdisciplinaire sur l'« interactivité » », *Communication et langages, 128* (1), 97–110. doi:https://doi.org/10.3406/colan.2001.3079.
Norman, D. A. (2002). *The design of everyday things*. New York: Basic Books.
Pachet, F. (2006a). *Enhancing individual creativity with interactive musical reflective systems, musical creativity*. Hove: Psychology Press.
Pachet, F. (2006b). Interactions réflexives. In Y. Orlarey (Ed.), *Actes des rencontres musicales pluridiscipinaires*. Lyon: Grame.

Pearson, R. E., & Simpson, P. (2001). From the entry for 'Reflexivity'. In: *Critical dictionary of film and television theory* (pp. 377–378). London: Routledge.
Phay-Vakalis, S. (2001). *Le Miroir in l'Art, de Manet à Richter*. Toulouse: L'Harmattan.
Quintilian. (2002). *The Orator's Education*. Vol. I: Books 1–2. (trans. Russell, D. A.). Loeb Classical Library.
Rheingold, H. (2000). *The virtual community: Homesteading on the electronic frontier* (revised edition). (Cambridge, MA: The MIT Press.
Schaeffer, J.-M. (1989). *Qu'est-ce qu'un genre littéraire?* Paris: Seuil.
Schon, D. A. (1984). *The reflective practitioner: How professionals think in action* (1st ed.). New York: Basic Books.
Schon, D. A., & Wiggins, G. (1992). Kinds of seeing and their functions in designing. *Design Studies, 13*(2), 135–156.
Sengers, P., & Gaver, B. (2006), Staying open to interpretation: Engaging multiple meanings in design and evaluation. In *Proceedings of the 6th conference on designing interactive systems*, pp. 99–108.
Sengers, P., et al. (2005). Reflective design. In *Proceedings of the 4th decennial conference on critical computing: Between sense and sensibility*, pp. 49–58.
Simon, H. A. (1996). *The sciences of the artificial* (3rd ed.p. 22). Cambridge, MA: The MIT Press.
Simpson, J. A., & Weiner, E. S. C. (2002). *The Oxford English dictionary*. Oxford: Clarendon Press.
Stam, R. (1992). *Reflexivity in film and culture: From don Quixote to Jean-Luc Godard*. New York: Columbia University Press.
Star, S. L. (1999). The ethnography of infrastructure. *American Behavioral Scientist, 43*, 377–391. https://doi.org/10.1177/00027649921955326.
Star, S. L., & Ruhleder, K. (2001). Steps toward an ecology of infrastructure: Design and access for large information spaces1. In J. Yates & J. Van Maanen (Eds.), *Information technology and organizational transformation: History, rhetoric, and practice* (p. 305).
Suchman, L. A. (1987). *Plans and situated actions: The problem of human-machine communication* (2nd ed.). Cambridge, MA: Cambridge University Press.
Todorov, T. (1990). *Genres in discourse* (trans: Porter, C.). Cambridge: Cambridge University Press.
Turkle, S. (2005), *The second self: Computers and the human Spirit* (Twentieth Anniversary Edition). (Cambridge, MA: The MIT Press.
Virilio, P. (1986). *Speed and Politics: An Essay on Dromology*. New York: Semiotext(e).
Witkowski, N. (Ed.). (2001). *Dictionnaire culturel des sciences*. Paris: Seuil.

Chapter 3
Creative Figures of Users

3.1 Introduction

In the late '80s and early '90s a number of intellectuals heralded better times, like the philosopher Pierre Lévy,[1] or prophesized disasters to come, like Virilio.[2] The origin of these predictions was legitimate enough – a concern for the impact of new technologies on our everyday life. Work, leisure, politics and sociability seemed to take on new dimensions and people were either afraid that they could lose some valuable aspect of their lives, or excited by the new opportunities created by information technologies. However, the task at hand for social scientists was to assess what really changed and how information technologies (IT) actually reshaped our lives. In particular they were worried that the optimistic vocabulary used by engineers and marketers to describe these new technologies could mislead the users and confuse their understanding of these new media. Hence the opening of a field of enquiry that has definitely contributed to a better understanding of these technologies in the middle of fierce media controversies. This research field used social, semiotic and ethnographic analyses and stood its ground between inflated marketing claims and overly pessimistic prophecies. On a more local plane, in my school, communication sciences researchers and information technology engineers often disagreed over the names of artifacts. For communication scientists,[3] words like "interactivity", "page", "navigator", produced and used by IT scientists were judged to be ill founded metaphors. They would provide alternative metaphors (for instance "signe passeur" – "go-between sign"- instead of hyperlink), to account for new media,[4] how they work, how they make sense, how people use them. I came to fully endorse their methods of enquiry and analysis.

[1] Lévy (1994).

[2] Virilio (1994).

[3] Souchier et al. (2003).

[4] Manovich (2002).

A. Gentes, *The In-Discipline of Design*, Design Research Foundations,
DOI 10.1007/978-3-319-65984-8_3

However, my work took a different turn based on the assumption that some common ground could be found that would benefit both engineering sciences and humanities. The goal was to find not only a common subject of interest, but also a context where our skills and concepts could be useful to each other. Figuring out what research in engineering meant was the first step. I therefore started by analyzing the institutional frameworks of engineering research.

3.1.1 Who Cares About the User? First Steps of a Survey

Everybody seems to care about the user: ethnographers, anthropologists, psychologists, management scientists, and economists. Each of these disciplines brings a point of view on what human beings or groups are and how they can be studied. Social scientist's expertise has been used by engineering teams to contribute to the emergence of IT. However, involving social scientists in a design process is still a challenge and a lot of literature explores the different ways that engineering or design projects can include different user-centered interests. Many will point out the differences in languages and cultures, the natural defiance between disciplines, and power struggles. In a best case scenario, social scientists are involved at the beginning of the ideation and during the development process; in a worst case scenario, they only intervene at the end of the project to test the technology and somehow legitimize it. In addition to the social and political reasons that explain why disciplines have a hard time collaborating, the main difficulty, in my opinion, is epistemological. Bringing in these disciplines, like social sciences or humanities, that study what the world is, to contribute to what the world could be, is a deviation of their initial purpose. Social sciences were not designed to justify a technology or to augment the world with new artifacts. Social sciences analyze how technologies are deployed and used as part of a larger range of activities that societies find meaningful: labor, leisure, urbanism, family, etc. This in itself is legitimate enough. They have a goal of their own and to think that these goals can align with invention needs to be carefully examined.

However, computer systems have a huge impact on activities, communication, culture and even the nature of knowledge itself. It is therefore more important than ever to explore how social sciences and the humanities can actually be incorporated in a research or development project and to see what changes are wrought both for the epistemology of engineering sciences and social sciences.

One of the ways to examine this process is to be involved in engineering teams and observe what happens. Rather than imposing a set of rules from the beginning, my idea – close to that of other social scientists' like Sander in the '90s for example[5] – was to understand what engineering scientists expected from social scientists. As it happened, reactions and expectations were quite diverse. While engineering scientists mostly expected social scientists to be specialists of use, and somehow to

[5] Sanders (2002).

represent users, they soon acknowledged that notions such as aesthetics, art, but also media, and more generally the humanities were providing important insights and contributions to their design thinking.

To start with, I needed to better understand how the "user" is a concept that shapes technological research and design. The concept appeared at different levels of the research programs and projects. It was present at a macro level, that is to say at the policy-making level as an obligation and part of the value assessment. It was also operational at a micro level in scientists' everyday activities since it structured tests, interviews, and state of the arts that scientific teams did. Bringing these two levels together has proved to be an obstacle in the analysis of science in the making. In my experience, it is difficult to articulate micro analyses of interactions between actors, semiotic analyses of their scientific texts, indicators of research and development, public policies that foster the development of information technology, and the society of information (Castells, Cardoso[6]). However, the illusion that those worlds do not cross is denounced by Latour in "Give me a laboratory and I will raise the world".[7] He shows that laboratories and the world of policy makers can be aligned and that the strategy (for instance that of Pasteur in Latour's demonstration) is to enroll other groups into the scientific argument. Though laboratories are not always the winners of the political battle as pointed out by Pam Scott,[8] the purpose of this chapter is to show that a whole rhetoric of research policy is not contradicted but slowly redefined so that each discipline carries out the purpose of the project in its own terms. The careful analysis of discourses and activities shows that macro and micro levels are indeed interrelated and, more importantly, that the figure of the "user" plays a special role in this alignment of goals and actors. Indeed, the apparent unity of a "user" hides a diversity of profiles – real and fictional – and divergent goals that boost the dynamics of invention.

This chapter therefore describes who takes care of the user at the level of institutions, at the level of interactions between actors, and finally within every new media. Three main questions are studied.

- Who takes care of the user? For many years now, engineering research frameworks have advocated that projects include different disciplines that address users' needs. While this goal seems worthy of interest, it is also defined in such a way that it could block innovation. This part of the chapter looks at the ruses/strategies that are deployed by research teams to avoid the pitfalls of the system, and in particular, how disciplines finally adjust to overcome their different views of the user.
- Who is the user anyway? Multidisciplinary research teams face the challenge of defining what is their common object of research, while different viewpoints on technologies – and subsequently users – are present. Convergence is hard to obtain because each discipline fights for its own conception of the user. This part

[6] Castells et al. (2006).

[7] Latour (1983).

[8] Scott (1991).

of the chapter describes what shifts occurred in a project that made it possible to share a "figure" of the user.

- The last part of this chapter then describes the facets that need to be taken into consideration while defining the user. I suggest that the "metaphor of the user" contains three figures: multi-tasker, aesthete and reflective practitioner. I use the word "figure" because what is produced by design teams is neither the real user nor is it a purely fictional character: in between reality and fiction, in between social sciences and the humanities, this figure is both an indirect description of real people and a speculative character.

3.1.2 Politics of Prudence: "Know Thy Man"

Should we really meet a user at the moment of invention? It has been highly publicized that Steve Jobs repeatedly refused any kind of human factors research prior to the invention of his artifacts, on the basis that users cannot anticipate something that is yet to come. The involvement of "users" in creating a vision of what they might be doing in the future on the basis of what they do today has thus been challenged. However, including users (in fact potential customers) is part of a strategy that I like to call "industrial prudence". Industrial prudence includes an obligation to anticipate the future of the technology in society: its impact on activities, its potential economic value, its consequences in terms of labor. As analyzed by Akrich, the technical innovation is part of an environment and the role of users is primordial in "aligning the object to the context in which it should be integrated". (1989) In other words, the industry is careful to anticipate as much as possible, the outcomes of its research especially in terms of markets. The massive trend towards open innovation,[9,10,11,12] is of course part of his trend. As studied by Von Hippel,[13] the user's contribution can be decisive in the innovation process to adjust and boost the offer.

One of the main questions remains: how to integrate a concern for potential users while inventing these technologies. Different fields of engineering were urged to involve users to avoid bad products. Nigel Cross, in 1972, painted a very pessimistic portrait of designers. He put forward their lack of concern for the consequences of their products. "Professional designers in every field have failed in their assumed responsibility to predict and to design-out the adverse effects of their projects. These harmful side effects can no longer be tolerated and regarded as inevitable if we are to survive the future … There is certainly a need for new approaches to design if we are to arrest the escalating problems of the man-made world and citizen participation in decision making could possibly provide a necessary reorientation[14]". (p. 11)

[9] Prahalad and Ramaswamy (2004).

[10] Ritzer and Jurgenson (2010).

[11] Zwick et al. (2008).

[12] Xie et al. (2007).

[13] Von Hippel (2006).

[14] Nigel Cross and Design Research Society (1972).

Quoting Panu Korhonen of Nokia, Liam Bannon recalls the evolution of users' involvement in Human Computer Interface. "In the early days the Nokia HCI people were told: "Please evaluate our user interface, and make it easy to use." That gave way to "Please help us design this user interface so that it is easy to use." That, in turn, led to a request: "Please help us find what the users really need so that we know how to design this user interface." And now, the engineers are pleading with us: "Look at this area of life, and find us something interesting!".[15] In this summary of user-centered interests, we can read four types of involvement of social sciences within engineering research.

The first stage consisted of testing the artifact after the invention. Changes could only be minor because the structure, the functions, and the architecture of the technology, were no longer questioned at that point. The product to be tested was like a black box without direct and visible correlation between the aspect of the machine and its use. In semiotic terms, it means that there is no indexical relation between the thing and the sign, causality between the tangible and the sense, that people could read. In that context, the user is only a tester who has to adapt to the new artifact. Social sciences are used in a very limited sense, close to clinical psychology. They set up tests with users in labs and assess, after the fact, that the technology is not too hard on the users.

Social scientists can also be involved at the stage of the invention. Their contribution consists of anticipating what the users might find useful and easy to use. It is a major shift because social scientists now intervene at an earlier stage of the project and not simply at the end. They have to develop a model of what is going to "work" for the user. First, they look at "natural" interactions. Then, the assumption is that they can use their observations to come up with specifications for new technologies that are going to mimic or accommodate human activities.

Thirdly, social scientists are supposed to be able to understand people's needs. Elaborating on Maslow's pyramid of needs,[16] they can come up with recommendations to address people's needs thanks to the technical system. Researchers in engineering can then adjust their invention to these needs. This is also a major twist in the epistemological finality of social sciences, since they no longer analyze what people do, but rather extrapolate what people could do.

Finally, social scientists are turned into scouts, who scan whole areas of activities to figure out what is missing and suggest new products to engineering scientists. It is not only a huge broadening of the field, it is also a considerable shift in what social sciences can do, from observing what people do to finding what is not "there" to fill up a "supposed" need. Their contribution is no longer about trimming a technology to make it acceptable. They have to find new spaces for technological inputs. From adjusting the invention to what already exists, to finding new ideas, the growing focus on uses and users is defined both as a bottom up innovation method and as a guarantee for adjustment. The expectation is that social sciences will reduce the difficulty and time to turn an invention into an accepted innovation.

[15] Bannon (2011).

[16] Maslow (1943).

More quickly said than done! In the next section, I point out some aspects of research infrastructures where the articulation between social and engineering sciences is being promoted and financially supported. My focus is on the discourses and how they present pluridisciplinarity in engineering invention and examining their underlying epistemological assumptions. In the second section, I will concentrate on the actual confrontation of disciplines and how they have to evolve to achieve their task of invention with a certain idea of the user in mind. Who is the right user for the invention? What role does the user play in the interplay of disciplines?

The following section will then discuss what the "user" means and how the "figure of the user" is a pivotal concept, almost a philosophical metaphor, that helps bring disciplines together.

3.1.3 Aesthetics: Answers Are Not Found in Numbers

To introduce the notion of "figure" of the user, I would like to first point out the major difference between social sciences and the humanities that can be summed up by this anecdote that could be called: user-centered painting.

It is the story of two Russian emigrant artists Vitaly Komar and Alex Melamid who, in 1994, decided that they would produce a work of art along marketing principles.[17] Thanks to the Dia Center for Art in New York and with the support of the Nation Institute, they were able to hire Marttila & Kiley, Inc. to conduct the first poll. The marketing company sent questionnaires to people about their taste and expectations in terms of aesthetics and art. Questions were such as:

> Favorite color? Prefer modern or traditional art? Preferred type of art? Art to fit decor or art you like? Prefer older or newer objects for home? Prefer wild animals or domestic? Prefer outdoor or indoor scenes? Prefer religious or non-religious theme? Prefer sharp angles or soft curves? Prefer geometric or random uneven patterns? Prefer expressive brush-strokes or smooth canvas?.

The same poll was later conducted in more than a dozen countries, like China, Russia, France, Brazil, etc. In 1994, they began the process which resulted in *America's Most Wanted* and *America's Least Wanted* paintings, which were exhibited in New York at the Alternative Museum under the title "People's Choice."

The artists' interest in such a process is two sided: first they wonder what would art look like if it were to please the greatest number of people? They are also concerned by the industrialization of culture and question the future of a society that is governed by opinion polls and market research. Though art is not *a priori* affected by this tendency, they push the logic to its extreme so that they produce an art in which we can believe because it is based on statistics. This faith in numbers is not

[17] http://awp.diaart.org/km/intro.html

only a characteristic of contemporary societies, it is also an undercurrent of art. The artist Alex Melamid noticed in particular that:

> *In a way it was a traditional idea, because a faith in numbers is fundamental to people, starting with Plato's idea of a world which is based on numbers. In ancient Greece, when sculptors wanted to create an ideal human body they measured the most beautiful men and women and then made an average measurement, and that's how they described the ideal of beauty and how the most beautiful sculpture was created*[18].

What is quite striking in this process is the result: most countries want a peaceful outdoor landscape, with a small group of people, and with blue as the major color. In other words, and as the two paintings show (Figs. 3.2 and 3.3), if based on polls, art of the twentieth century would look like a painting of the eighteenth century such as Fragonard's (Fig. 3.1).

Here goes innovation in painting, if it were user-centered!

Of course, what is interesting in this artistic process is not only the result but the fact that we are suddenly confronted with how art makes sense and how it relates to audiences. This conceptual art work is analyzed by Andrew Ross.[19] For the art critic, the artists are not only satirical about society, they deeply question what art is as a social and political category. In between official or state art, fine art, or popular art, they explore the separation of the different art worlds. I think that this experience in art is the epitome of what creation is about and a good starting point to think about authorship and how people can be involved in the process of creating, as well as what it means to produce a meaningful artwork or design product.

The first question that we should ask ourselves when looking at these rather drab and totally non original pictures is whether the question of creation governed by numbers is relevant to creation. Obviously if everybody wants the same thing, there are two consequences. There is no innovation: creators are stuck with a cultural representation that is legitimate in itself but reflects more a status quo than a progress towards new forms, new ways of looking at things, new viewpoints. When I say new, it is not only a question of novelty for avant garde's sake. It is rather that the art world questions our relation to what life means at a certain point in time and space. Contemporary life is not about lakes and bears or George Washington even if nobody contests the fact that these images are part of the culture of these countries. The task of the artist is to engage herself and spectators in a contemporary adventure, not a remembrance journey. Art when it is created is always contemporary and as such it might shock, bewilder, surprise, because it reconsiders the previous regime of creation. More to the point, art says something about the status of men in their contemporary world. Works of art that are considered relevant[20] say something true about their time.

But from the viewpoint of a market that is focused on maximizing profits, true art risks something very important: it risks having very little audience at first. The

[18] http://awp.diaart.org/km/intro.html

[19] *ArtForum*, January 1995, pp. 72–77, 109.

[20] Rochlitz (1998).

Fig. 3.1 Jean Honoré Fragonard/A Game of Horse and Rider/c. 1775/1780," – French Paintings of the Eighteenth Century, NGA Online Editions, https://purl.org/nga/collection/artobject/32683 (accessed September 19, 2017)

history of art is full of these clashes between established art, that is to say art that people are used to and "contemporary" art that offers a new way of looking at things and that is consequently considered scandalous. People who produce new technologies, new media, new applications, face exactly the same dilemma. What measure of risk is a society, an industry, or a public laboratory ready to take? What are the arguments for limiting the risks? What do they do to limit the risks of not finding any audience?

While these questions are legitimate and address the management, and economics of innovation, this book is about designers' practices when they take risks and deal with innovation without marketing it first. The idea is not to exclude users from the design project but to better understand and qualify what role the "figures of the user" play in the invention process. If the user who helps the invention is not the

Fig. 3.2 Vitaly Komar and Alex Melamid/"Most wanted picture – USA" 1995, http://awp.diaart.org/km/usa/most.html, (accessed September 19, 2017)

Fig. 3.3 Vitaly Komar and Alex Melamid/"Most wanted picture – Russia", 1995 http://awp.diaart.org/km/usa/most.html (accessed September 19, 2017)

user found in opinion polls, who is he/she? How do researchers and designers build a representation of the user that helps their creative skills?

3.2 Pluridisciplinarity in Research Frameworks: The Allegory of the User

A first figure of the user appears in texts organizing information technology research programs under the wing of pluri-disciplinarity. The European Union research framework and some programs of The National Science Foundation[21] aim at fostering multidisciplinary teams as well as at supporting diverse organizational representations.[22] An important claim favoring multidisciplinary collaborations is that they promote creativity and innovation but also that they meet "The User". Advancing the needs of the user is presented as the ultimate goal of this creativity. "The engineers look for "needs" that are not "satisfied" and pledge to build the appropriate machines. "Common threads in the different meanings of need seem to be, firstly, their reference to relationships between individuals, their goals, and the means of achieving these. Secondly, these relationships are characterized by a certain degree of necessity or urgency[23]". Science is allegedly done in the interest of society.

3.2.1 Disembodied Technologies: The Rise of the Ellipse

However, a rapid analysis of these texts organizing the framework of IT research since the new millennium shows a paradox: the emphasis on users' needs coincides with an emphasis on dematerialization. The implication is that it is a good thing if the user does not see the technology. As it would take too much space to quote all the founding texts, I will only present a few examples of this discourse.

Let us start with the metaphor: "ambient intelligence". It was used in workshops[24] organized by Philips in 1998, published in a Dutch magazine in 1999,[25] and was picked up by the European Commission to define future research goals.

> The workshops were aimed at developing different scenarios that would lead a high-volume consumer electronics industry from the current world, which was called fragmented with features, into a world near 2020 with fully integrated user-friendly devices supporting ubiquitous information, communication and entertainment.[26]

---

[21] See in particular the NSF's office of multidisciplinary activities.

[22] Cummings (2005).

[23] Keinonen (2010).

[24] Zelkha (1998).

[25] Aarts and Appelo (1999).

[26] de Ruyter (2003).

In particular, the home and its artifacts were the focus of the extremely popular Philips' "homelab". Experiments turned traditional objects, home appliances and the house itself into new products: transforming TV into a more explicit and perceived collective experience, turning photo albums into shared content displays, etc. In the texts posted by Brian Epstein,[27] who participated in Philips' workshops, the focus is on user-friendly artifacts. In a "Power point" on the "Digital Living Room" that presented the results of these experiments in 1998, it was stated that the: "Single most important movement today in all areas of computing is human-centric".[28]

This human-centric focus is nonetheless based on an ellipse: the disappearance of the tangible artifacts. In other words, these technologies are not considered media but tools or mechanisms that are going to fade into the fabric of everyday life and activities, an ideology that is shared by many as shown and criticized by Blum for example.[29] Anthropologists are therefore needed to see how "Life" in the home is organized and to figure out how computerized technologies are going to support activities but, strangely enough, without thinking about it, or visualizing the technology or people struggling with it for that matter. Along with the elliptic technology, the Allegory of the User slowly appears.

For instance, researchers would be presenting Ambient Intelligence's benefit in very strange terms as in the following excerpt of a power point presentation: people should use them "unconsciously" (Fig. 3.4).

This presentation of 1998 is representative of the general discourse of the time, but if one steps back and thinks about it, it seems strange that people should be empowered when they are unconscious. Other descriptions build on this idea by advocating the invisibility of these technologies.

> Ambient intelligent environments are ones in which the user is surrounded by a multitude of interconnected embedded systems, **which are invisible and non-invasive**. These ubiquitous systems are able to locate the user and take into account the context of its demands to understand its intentions, learn from its behavior and adapt itself to him. In such an environment, man-machine interactions is obviously **natural**… and speech technologies pervasive.

The less we see these technologies, the more elliptic they are, the more natural they are argued to be, and the more they are able to serve the Allegory of the empowered and fully satisfied User since his/her every need is anticipated and satisfied.

The European Commission decided to base its 6th research framework (2000) on the concept of Ambient Intelligence. In 2001, a working group of the highly influential Information Society Technologies Advisory Group, (ISTAG) chaired by Dr. Martin Schuurmans (CEO of Philips Industrial Research) wrote a report with a series of scenarios to explore the future of connected and embedded systems. They, of course, focused on users: "The concept of Ambient Intelligence (AmI) provides a vision of the Information Society where the emphasis is on greater **user-**

---

[27] https://epstein.org/ambient-intelligence/

[28] Draft of the "Digital Living Room" presentation by Roel Pieper circulated within Philips, 24 June 1998 retrieved April 2016 https://epstein.org/ambient-intelligence/

[29] Blum (2013).

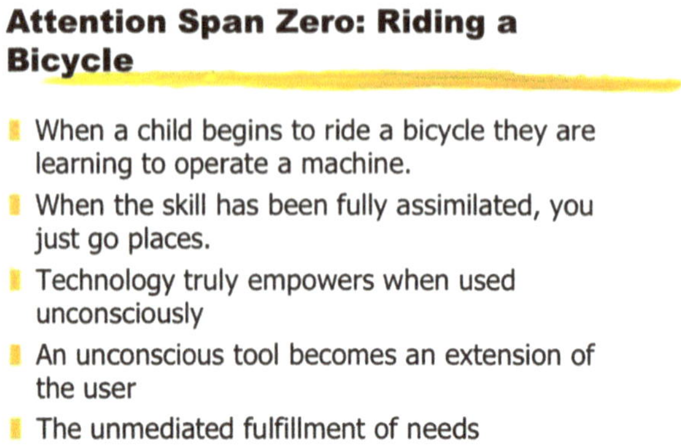

Fig. 3.4 24 June 1998 – Draft of the "Digital Living Room" presentation by Roel Pieper circulated within Philips, retrieved April 2016 https://epstein.org/ambient-intelligence/

friendliness, more efficient services support, **user-empowerment**, and support for **human interactions**[30]".

IBM developed its own trend of ubiquitous computing through the expression: "pervasive computing" that is also widely used to this day, though with a slightly different perspective from the pervasive metaphor. Coming from distributed systems and mobile computing, it emphasized the infrastructure that large organizations might create to support the exponential growth of connected objects (like in the Internet of Things). Pervasive computing emphasizes "**effective use** of smart spaces, **invisibility**, localized scalability, and masking uneven conditioning[31]". Again, invisibility is one of the focus points as well as "effective use". If I sum up the value proposition: when things are invisible and we act unconsciously, we are empowered, efficient, and utterly satisfied.

The common point of these different metaphors is that they tend to hide the tangible technical artifact. The emphasis is on "dematerialization" as a desirable state: "The process of dematerialization also affects the audiovisual and multimedia sector, from design and development to dissemination and adoption, use and standardization of solutions[32]". By centering their discourse on dematerialization, the various international programs contribute to an image of an immaterial technology. This figure of speech is often reinforced by the so-called "need" to hide the complexity for the end-user. This vocabulary blots out the materiality of these technologies and does not consider them as new media. While the argument makes sense on some level, the discourse impacts the way people think about these technologies in relation to their potential users. In particular, they do not put any emphasis on design features since the artifact is supposedly transparent. Engineering research is very much thought of in terms of abstract concepts, applied mathematics, applied

[30] Ducatel and Bélgica (2001).

[31] Satyanarayanan (2001).

[32] Project Call, RIAM 2006.

physics. Of course more "user-friendly" disciplines – like human computer interaction – will have a different discourse but they still have to cope with this general framework that constructs the Allegory of the User.

3.2.2 *Re-embodied Technologies: The Allegory of the User*

"Uses" and "users" fortuitously reincarnate these technologies that seemed to fade away. The EU funding process, for instance, criticizes research projects if they do not include enough consideration for what these technologies are going to be used for. National programs assess the results in the same way:

> LAN technologies including fixed and wireless services for mobile users is a priority this year for the RNRT. Several projects have been submitted on this subject under the 2002 call for projects. However, none has been fully satisfactory. Overall, these projects involved research and technical development of good level, but little reference to services and uses induced by these technologies, while it is a key issue for this type of systems.

More broadly speaking, international calls for proposals include a fair portion devoted to what I call "user attention", that is a way (in fact many) to include a reflection on what is to become of the technology in the hands of a user or, more broadly speaking, how the technology is going to be integrated in society.

Sociology or ethnography and other social sciences[33] appear to be the disciplines that can take care of this part of the research. Social sciences supposedly reinforce the scientific process by analyzing uses, and introducing and possibly informing methodologies "upstream". Their contribution to the design process has evolved from concepts such as "user-centered design[34]" that focuses on the users' perspectives, but does not necessarily include users as actors in the design process, to "participatory design[35]" where users take part proactively in the design,[36] or "collaborative design" that refers to the "creativity of designers and people not trained in design working together in the design development process", and blurs the frontier between designer and user.

However, if one reads carefully the calls for proposals, some theoretical shortcuts appear. There is admittedly no use for something that does not exist. More than anything else, there is no way that social scientists can infer from what they observe, a definition of a future technology. At the end of the project, the tests can rely on methods of observation that come from social sciences (either in the lab with usability experiments, or in the field with ethnographic methods) but they can hardly infer social rules while the whole process is totally experimental. The second argument is that users have needs. This argument is interesting because it gives a way to describe

[33] Ergonomy, Anthropology, and more generally what is alluded to by the broad term of "user studies".

[34] Norman and Draper (1986).

[35] Sanders, « From user-centered to participatory design approaches».

[36] Halskov and Hansen (2015).

the non existence of the activity and justify the invention of a technology that will support it. The presupposition is that there are either very local needs (like a community that has specific goals and activities that the research team has to know) or very universal needs (like boys meet girls in a Facebook way). At best, observing and questioning people help incremental innovation because each artifact can be tuned to specific contexts. But these "needs" do not explain radical innovations mostly because people cannot express needs that would so much depend on a different technology. As pointed out by the media specialist, Yves Jeanneret, the notion of use is instrumentalized by the notion of needs and the benefits that are granted through the use of the object.[37] In the paradigm of "use and gratifications",[38] social life provokes needs that can be gratified through diverse means, of which the use of media is one. "Their central notion is that mass communication is used by individuals to connect or sometimes disconnect) themselves – via instrumental, affective or integrative relations – with different kinds of others (self, family, nation, etc.). The scheme attempts to comprehend the whole range of individual gratifications of the many facets of the need "to be connected". And it finds empirical regularities in the preference for different media for different kind of connections (Katz, Blumler, Gurevitch[39]).

3.2.3 *First Conclusion: Frameworks for the Allegory of the User*

To sum up the situation, technologies are transparent, the user is a bunch of needs, and the use must be unconscious.

On a methodological level, the frameworks lack a vision of the actual artifact. They do not include disciplines that would look at the formal and cultural traits of the invention. Second, there are no guidelines on how to combine social sciences on the one hand, that have a quite complex vision of users, with engineering sciences, on the other hand. The underlying assumptions of these calls for proposals is that engineering sciences have something in common with fundamental sciences and social sciences. However, the overlap is not clear, nor the actual means to do it. While the frameworks give a general background, and more importantly the means to do so, they are not sufficient to define the precise interplay of disciplines engaged at the time of the invention and how this interplay involves different figures of the user.

A closer observation of research practices in these projects can help us understand how multidisciplinary teams can resolve the tension but also reframe the question and offer new ways of weaving social and engineering sciences together. My

[37] Jeanneret (2007a).

[38] Katz et al. (1973).

[39] Ibid.

hypothesis is that the figure of the user is more than just an accessory to engineering sciences, or a political tool to legitimize technical research. It is a pivot point in the articulation of disciplines and their evolution.

3.3 Who Is the User? How to Organize Different Viewpoints on the Subject

Everybody is interested in the user. The question is who is the user anyway? Every discipline has a special way to define the "user". Social sciences have a rich language of users as workers, families, dependents, communities, etc. This language is born out of the observation of significant social behaviors with ethical, political and social implications which help isolate and qualify a group of people. Social scientists consider how these groups define concepts like suicide (Durkheim[40]) or drug addiction (Howard Becker[41]) and how people define themselves, their values, their activities.[42] Who is the "user" from a social science point of view is therefore not such an easy question to answer.

Defining "The User" was one of the main challenges of a project that we had with the Museum of Arts and Crafts in Paris. The project could be seen as a brainteaser: how do we build an application that is based on a distributed mobile infrastructure, for museum visitors to interact and learn and play, while we do research. The Museum imposed a complex set of constraints: (1) the game should be both educational and entertaining, (2) it should use a limited infrastructure – in other words no Wi-Fi network, and depend on limited operating costs – no UMTS/HSDPA network to transmit data – (3) it could be played by anyone, whatever age or social category, and preferably by a family, (4) it could be played several times by the same player. The game designers were quite desperate because they legitimately thought that the whole sets of instructions would never provide any fun at all.

The following part of this chapter describes the different starting points of the actors and the paradigm shifts around the figure of the user that finally led to the game that was successfully designed: "PLUG: the secrets of the Museum".

[40] Durkheim (1997).

[41] Becker (1966).

[42] Berthelot (2001).

3.3.1 PLUG: The Multifaceted User[43]

The PLUG project[44] took place within the RIAM program (Research and Innovation in Audiovisual and Multimedia) of the French National Research Agency (ANR). Like many of its kind, this institutional framework strongly advocates a convergence between information and communication technology (ICT) and economics and social sciences. The ANR wants to support breakthroughs in engineering research based on computer science. However, it also poses as a condition that the project specifies how these results can be socially and economically successful. The primary focus of these programs is therefore to transform emerging techniques into social realities. To that effect, actors deploy strategies to develop the technical object both as a scientific and social product. But project participants are also representatives of institutions (the museum, public labs, and the video games company) and research areas (information technology, design, sociology, information and communication sciences). And of course, with each discipline comes a different vision of the user.

While the institutional framework encourages the collaboration of these different disciplines, it is not enough to actually weave the diverse research components: the participants also need to have an idea on how to collaborate effectively. They do not have a recipe.[45]

The team included the French Museum of Arts and Crafts in Paris (Musée des Arts et Métiers), a game design company: Tetraedge, a French telecommunication operator: Orange, and academics in pervasive computing and social sciences from the Conservatoire National des Arts et Métiers (CNAM) and Institut Télécom. Along with the engineering team, two researchers in information and communication sciences and one interaction designer participated in the creation, development, deployment and testing of the prototype of the game: "The Secrets of the Museum" that help people discover and learn about the museum artifacts. The social scientists had a double agenda: to contribute to the product, but also to study how the different actors were doing the research and designing the prototype. Documents, meeting notes, email exchanges, models, "deliverables" that accompanied the production were gathered and kept for that purpose.

One of the major findings of this experiment, was the realization that the participants altered the boundaries of their disciplines as well as questioned concepts that were usually associated with them. To design the new object, the actors had to challenge a number of their assumptions. Each discipline under-determined the others. These "paradigm shifts" happened as "a movement of expansion or complexification of the initial field of investigation, a process of exchange and transfer models; movement of reflexivity and self-analysis."[46] This situation is exemplary of what I

[43] This part of the chapter is using material presented in a few conferences. I wish to thank the co-authors for their authorization to reuse this material.

[44] Simatic et al. (2009).

[45] Cummings (2005).

[46] Berthelot, *Epistémologie des sciences sociales*. p. 226.

Principal actors	Basic user vision
Museum	The visitor as heir of the nation
Game design	The gamer
Information and Communication Sciences	The interpreter of information
Engineering	The manipulator of technologies

Fig. 3.5 PLUG. Disciplines and main user figures

call the in-discipline of design that I fully develop in Chap. 7. Right now I want to focus on the different figures of the user.

The following table is a simplified view of how stakeholders represented the user at the start of the project. It sums up the main direction of their vision and shows how diverse the vision was from the very start (Fig. 3.5).

A museum has a mission: to protect, analyze and promote the cultural heritage of the Nation. Even when we looked at other recent RFID experiments in museum, a rather traditional view of the visitor emerged: someone who is expecting to learn something from the museum and who accesses the information in the orderly fashion that is offered by the curator. Pervasive technologies present an interesting potential for museums because they can change two characteristics of the visit[47]: mobility and interaction with the artifacts. However, looking at the actual pervasive applications, many[48,49] presupposed the same type of mobility for all visitors: a linear progression from one artifact to the other, and, at each stop, information delivered in "audio-guide fashion", with top down communication.

The vision of the game designer is of course very different. To start with, users might be more interested in indulging in competitive gameplay than in actually learning about the material in the museum. As game players they might need to be challenged with an increasing level of difficulty. Another potential challenge for game designers is that a museum might have several thousand people come through it in a day; the time that an individual should interact with a display must be limited to be successful too. Game designers have to answer questions such as: "What are the obstacles that I will put in their path?", "What is the purpose of the game: solving a puzzle, beating an opponent?", "How will the player win the game?". While the user of the museum seeks to belong to a cultural community, the gamer may want to belong to a community of other gamers who simply have fun.

For the information and communication researchers, the user is primarily an interpreter of information and a communicating person. One of the main questions is how to provide the right information in the right format through the right media, to the right person. The focus on users is on how to engage their curiosity, learning and interpretive skills, and on how to offer proper cultural mediation. Soon they had to consider the specificity of the Museum of Arts and Crafts, which is a heritage

[47] Gentès and Jutant (2009).

[48] Hsi (2002).

[49] Le Marec (2012).

museum of history of technology, in between two paradigms: the museum of science versus the art museum. Visitors can manipulate some devices but mostly mediators offer demonstrations at fixed times in theaters and the majority of artifacts are formally exhibited.[50] Display cases and pedestals keep visitors at a respectful distance from the objects. Cartels do not deliver much information about the workings or use of the object. The information is therefore staged to inspire respect, keep at a distance, produce relevant historical information but no practical information (how did people use these artifacts), or mechanical and engineering insight (how do these artifacts work).

Finally, engineers primarily consider the user as someone who manipulates a device and activates a system that supports the activity. The users can accomplish tasks thanks to the device. Engineering scientists focus on the ease of use and accessibility of the device.

What did we learn from this first list of figures of the user?

- First, that the same person can and will be defined by a multitude of characteristics.
- Second, that we can hardly speak of needs to sum up all the characteristics. The situation builds a multifaceted figure of the user and activities, literacies, emotions are co-constructed by the different institutions that the user belongs to or inhabits for a while.
- Finally, the Allegory of the User as a bunch of needs empowered through transparency seemed to be far from everybody's representation.

The challenge is also that all these characteristics have to somehow crystallize in one activity/situation supported by one device.

But more importantly, such a device already exists, since audio-guides are common devices of museums and downloading information about a specific item is a banal internet feature! The invention is hard because all the "needs" as well as the delicate ritual of museum visits are already taken care of by similar devices/technologies. However, it was precisely the diversity of user's definitions that forced every discipline to reconsider its position and concepts and come up with a new artifact.

3.3.2 From the Non-descript User to a Complex Visitor

Each discipline in the project had a rich and complex theory of the user. But each discipline on its own could not encompass the whole figure of the user. At first, it seemed that they could use a minimal figure of the user- "the smallest common denominator user"-, or that they could be all put together and have a comprehensive view of the user – "the additive user". In fact, each definition challenged the others. The following table shows how the perspectives changed. The cases show the

[50] Davallon (2000).

	Museum	Game design	Information & Communication Sciences	Engineering
Museum	Museum visitor	What is a playful relation to the museum?	What is the new role of the visitor in the communication process?	How to define a museum technology?
Game design	How can a visitor of a museum institution be a player too?	Gamer	What is a game that is a strategy of communication too?	How can the device be part of the physical architecture of the place?
Information & Communication Sciences (ICS)	What are the means to observe and design visitors' strategies?	How do users look for narratives/ create narratives	Interpreter of information	What are the mediations offered by technologies?
Engineering	What are the new gestures with the device in the museum?	How can contents be shared between the different actors?	What kind of media can be used to support pervasive computing?	Manipulator of technologies

Fig. 3.6 PLUG – questions between disciplines (under-determination)

questions that each discipline submitted to the others. The grey cases reflect the self-referential definition that is given by the discipline (Fig. 3.6).

For the curators, under the pressure of the game designers, the question was how to admit a playful relationship with the Museum while retaining the trust of visitors. Introducing the game and the technology triggered a reflection on the missions and modalities of museum visits. The question was also raised from an information perspective: how could the visitor participate in the communication and interactions between the museum and its visitors and also communicate and interact with other visitors? Finally, with the engineering team they looked into their tradition of technical devices that accompany the visitors and how to come up with a really original one (why make another audio-guide when you already have audio-guides?).

From the museum perspective, the gamer was not only someone in a material environment. The visitor has rights and duties towards the institution. The museum has a relationship of trust with its visitors who expect the museum to produce reliable content about the collections.[51] Game designers had therefore to take this into

[51] Marec, *Publics et musées*.

consideration. They also had to realize that the game and gamers were necessarily part and parcel of the strategy of communication of the institution. How could a game be defined in this situation where so much was at stake in terms of mediation of artifacts and knowledge? They also had to consider the device in relation to the specific architecture of the place and its other artifacts[52,,53] The fact that phones are prohibited in most museums had to be taken into consideration.

The researchers in information and communication sciences considered the Museum's communication strategy and the way it had to be embedded in the new message provided by the game design. Their discussions with the game designers also slightly changed their viewpoint from social sciences to the humanities. What kind of public narratives were already there and how to build on them to create a powerful narrative for the game? They also had to look into ways the RFID technology could change a whole set of uses of the mobile phone: from private to public use of phones, from downloading and receiving to creating, from indifferent to the environment to environment dependent. Finally the biggest change was that they had to participate in the creation of the new device, therefore switching from a position of neutral observers to a position of active conceptors.

Engineers also had to shift gears, since they had to consider the cultural impact of a new technology in a hundred years old institution. The Museum is not the laboratory nor the workplace. Suddenly the implications in terms of moves and gestures were foregrounded by the situated experiment. How could they deploy a technology that would not endanger the first mission of the Museum, that is to say the preservation of the collection? The second change was to consider data as contents. While engineers needed a clear vision of what data (quantity and type) could be available, game designers and social scientists were more interested in how it would make sense. Even the technical properties of the artifact had to make sense for the game and therefore technical characteristics were turned into properties of the game play. The idea of contents also pushed the participants of the project to consider themselves as researchers but also as authors (a point we elaborate in Chap. 4). Finally, engineers also agreed on a change of perspective when they adopted the word "media" rather than information technology to speak of their prototype. Indeed, the switch caused them to focus not on the so-called transparence advocated by the research frameworks, but on the aesthetics of the object, its shape, aspect, and the way it was experienced. The concept of media therefore introduced the question of cultural reception. The game was not there for game sake. It served the purpose of bridging the worlds between people and the exhibits. Therefore, the technology was no longer some neutral tool but was part of a system of mediation.

Our experimentation showed us that the different facets of the "users" could not be reduced to their needs. Epistemologically speaking it meant that there could be no social or natural "reductionism". There was no technological determinism either. Working on the "user" meant that disciplines changed a number of their concepts: from user to spectator, from gesture to reading/writing/narrating, from RFID tag to

[52] Davidson et al. (2007).

[53] Montola (2005).

screen to media. This goes way beyond a mere change of vocabulary such as it is described by Louis Bucciarelli, in his book Designing Engineers.[54] More than a shared vocabulary, it was a redefinition of concepts.

3.3.3 The Users Challenge the Users

The testers themselves came up with new figures by describing their experience. For instance, all the participants described this interaction with the technical device as "writing" and "reading". This seriously challenged information and communication scholars as they had to redefine what reading and writing mean. A new figure of the user also appeared: the user as an author, first because the signs on the screen allowed them to understand how the technology worked, and second because they could actually change the sequence of the narrative within the game. While swapping cards, players were changing the way the others would access the contents. The narrative of the game therefore changed and switched from a game of collection to a game where players decided on the course of action and defined the relation to the museum institution. The players became narrators. This also meant that game designers would come up with a new definition of what a narrative is (Fig. 3.7).

3.3.4 The Role of the "Figure of the User" in the Invention

These shifts demonstrated that it was impossible to come up with a unified version of the user. And it actually was not a problem. On the contrary, bringing together diverse views, forced the designers to create new figures of the user. Far from simplifying the user by finding a common denominator, the disciplines changed their definition of the user by expanding their representation and their understanding, enriching their different figures of the user. There was not a convergence on the definition of a precise word but rather a reconfiguration of the attributes of the user that allowed to define new set of properties for the technology.

In particular, they considered the user as:

- Someone who is defined by the situation. During the creative process, the status of visitors, their expectations towards cultural education, the forms of legitimacy of the museum, were at the forefront of the design process. The profile of users became extremely complicated because contrary to design methods employing personas, the situation called for a combination of profiles: not only visitors but gamers engaged in competitive experiences and learners trying to take knowledge away from their visit.

[54] Bucciarelli (1996).

Fig. 3.7 PLUG. Transformation of focus

- Someone who plays with a number of material artifacts. This affected the way the technology was defined: what kind of device, how would it be distributed, how people already use it, what are the other devices that they use, how could this one be original compared to audio-guides?
- Someone who moves and employs "meaningful gestures" in space: how do people move in the museum space, where do they stop? What gestures are authorized as part of the interaction? How are they read as non threatening?
- Someone who is defined by his/her consumption of the media: what do visitors read? Where and when do they read it? How do users split their attention between the device and the actual artifacts? What genre of texts are they expecting? How do they recognize it?
- Someone who is defined by other actors: what kind of communication do visitors expect from the museum, or from each other? How do they manage their relation to knowledge in the museum space?
- Someone who learns from the device, the contents and others: how to create a self-explanatory device? How to rely on literacies to build new knowledge? How to reintroduce a social dimension to learning?

These definitions created a new way to circulate between different facets of the user and were no longer distributed along the lines of each discipline. Contrasting different figures of users redistributed partitions, creating new focus and divisions independently of disciplines.

In my opinion, this reshuffling of concepts – which does not mean that one discipline adopts the concept of another, or completes the definition of another, or sees the same concept from different disciplinary perspectives- is the key to understanding the in-discipline of design and radical innovation. The process of invention means that definitions change, they detach themselves from their disciplines as I will further develop in Chap. 7.

The user was not so much a target for the research project, or the allegory of the usefulness of the research, but a concept that forced each discipline to reconsider its own "figure" of the user. It was a generating lever that actually changed several concepts in each discipline.

3.4 Three Multidisciplinary Figures of the "User": Multi-tasker, Aesthete, Reflective Practitioner

Beyond the topicality of PLUG, there is a major lesson that we learned from this experience. We found three multidisciplinary figures of the "user" that are actually useful for the design of information technologies and applications: the user as an actor performing a task, the aesthete experiencing a sensorial and cultural relation to the object which has been addressed by other researchers like Nathan Crilly,[55] and the reflective practitioner representing the world and creating meaning with signs. Each figure is both and at the same time a relevant way to describe the user and a trigger to reconsider the stability of the definitions within the disciplines that represent that figure.

The table below is a summary of the different properties related to the different figures of the user (Fig. 3.8).

In the next section, I describe in more detail the different figures that can account for the "properties" of the user.

3.4.1 *The User as a Figure of Multi-tasker and the Designer as the Efficient Inventor*

The history of Human Computer Interaction is the history of how to define and measure user performance starting for instance with Steward Card, Tom Moran and Allen Newels' book: *the Psychology of Human Computer Interaction (1983).* The

1- figures of the « user »	Multi-tasker	Aesthete	Homo sapiens
2- our relation to the world	Action on the world	Passion (how we are affected by the world)	Representation, mediation of the world
3- figures of the designer	efficient inventor	virtuoso of norms	mediator
4- design focus	Activity	Taste - Emotion	Interpretation
5- models of the object	IT as a tool and prosthesis	IT as an artifact	IT as text, document, media
6- Design evaluation	Effectiveness, Efficiency, learnability	Pleasure / pain	Readability, communicability

Fig. 3.8 Figures of the user: Multitasker, Aesthete, Homo Sapiens

[55] Crilly (2010).

field has worked to understand how a user *might* include the user as a multi-tasker. The focus is on the interaction with the machine to accomplish a task. Donald Norman and Stephen Draper for example introduced the term user-centered design in the title of their 1986 book: *User-Centered System Design: New Perspectives on Human-Computer Interaction*. The article “Design for usability” by Gould and Lewis’ (1985) is also an often-mentioned early reference to User Centered Design (UCD) principles. Whatever the branch of user-centered design, we want to show that it is activity centric and that this has limited the approach to design.

In his article on the evolution of user-centered design, Keinonen shows the growing field of user-centered methods:

> These include human factors and ergonomics, participatory design (Greenbaum & Kyng, 1991; Schuler & Namioka, 1993; Sundblad 2009), human-centered design processes (Beyer & Holtzblatt, 1998; ISO 13407, 1999), usability measurements and inspections, i.e., usability engineering (ISO 9241-11, 1998; Nielsen, 1993), and design for user experience (Khalid, 2006; Koskinen, Battarbee, & Mattelmäki, 2003; Norman, 2003; Jordan, 2002). More recently, approaches such as service oriented design, transformation design (Burns, Cottam, Vanstone, & Winhall, 2006), lead user innovation (von Hippel, 2005) and worth-centered design (Cockton, 2006, 2008a) are expanding the umbrella even further.

In *Design Research through Practice*, Ilpo Koskinen points out that these methods have evolved from laboratory to field. First, the user was observed during her interaction with the machine out of context. Such criteria as speed, accuracy, primarily physical and psychological properties could be tested. But this first method was criticized as it did not take into consideration the fact that people are always acting in context, at certain times and in certain locations, with specific constraints. Several critics also pointed out that the concept of activity is not so clear-cut: Wendy McKay[56] for instance remarks that the interaction of people and technology is *co-adaptive* (Mackay, 1990). “People both adapt to the technology and they actively adapt it for their own purposes. Thus, the problem is not static: the “same” technology is often very different in different environments”. Sociologists have also warned that the use of the technology cannot be considered as solely an individual task. People are influenced by how others around them interact with the technology.

In any event, the focus is always on activity. Usability Professionals’ Association [UPA] (2008) defines UCD as “an approach to design that grounds the process in information about the people who will **use** the product. UCD processes focus on users through the planning, design and development of a product.” Two goals are targeted: in the tradition of ergonomics, the safety and comfort of the user; in the tradition of engineering, the efficiency of the activity in relation to goals. The machine is therefore designed to accomplish tasks. Its functions augment the capacity of the human body to act on the world. The challenge for the designer is as Redstrom emphasizes to have “‘definitions of use through design’ and ‘definitions

---

[56] Mackay Fayard (1997).

of use through use".[57] In other words, to introduce ways of testing new uses before they become "uses".

3.4.2 *The User as the Figure of the Aesthete and the Designer as a Virtuoso of Norms*

A strong turn in HCI also tends now to involve not only cognitive processes and functions but the emotion of the user[58,59] as a parameter to be taken into account when designing and evaluating design. In philosophy, the concept of action is paired with the concept of passion and I think that this larger focus helps us understand what is at stake when designers speak of emotions. In passion, the users are affected by the technology. They can be affected physically or emotionally. To generate or to control these emotions, designers develop aesthetic strategies. Aesthetics originally means the full experience that one can get from a situation or an artifact. While functionality is the prime concern of task-oriented design, aesthetics is at the core of experience-oriented design.[60]

Designers need to understand the users not only through their activity. They need to take into account the user as an aesthete. This has been a recent subject of interest in the HCI community.[61] In particular, McCarthy and Wright suggested that we need a more holistic view of the user.[62] As an aesthete, the "user" will look for styles and genres that she might relate to different forms of sensual experience. I like to take the classic example of music. In our occidental societies, music is still mostly based on Bach's "Well-Tempered Clavier" and major and minor are correlated to certain types of emotions: major is "heard" as energetic, optimistic, joyous, while minor is "heard" as reflective, nostalgic, even sad. There is, of course, no direct relation between the actual physicality of the sound and our emotions, but a long culture of associating these two modes effectively triggers these emotions.

There are at least two ways of dealing with the aesthetic experience. On the one hand, HCI and ergonomics researchers develop sensors and observation methods that evaluate the emotional response of the users. On the other hand, cultural studies and semiotics are necessary to assess the literacy, genre, style, the formal aspects of artifacts, more broadly speaking the aesthetics as it is defined by the humanities. Understanding the designer's activity is therefore to recognize that it is not limited to adjusting the systems to an activity but that it is also defined by the ability to play with norms, whether moral, social, or aesthetic.

---

[57] Redström (2008).

[58] Jordan (2000).

[59] Koskinen and Mattelmaki (2003).

[60] Crilly (2010).

[61] Boehner et al. (2008).

[62] McCarthy and Wright (2007).

3.4.3 *The User as an Interpreter, and the Designer as a Messenger*

The user of information technologies is not only a multi-tasker and an aesthete. As studied by Jeanneret,[63] media are objects that redistribute the relationship between the text (objects to be interpreted), the inscription (sign of legitimacy and memory), the dialog (opportunity to elaborate and confront thought) and authority (organization of power relationships). Contrary to face-to-face conversations, media organize a relation between absence and presence, an inscription of time in space. The material aspect makes it possible for symbolic changes to happen. The main concern is therefore not so much action or passion (or aesthetics) but representation and interpretation and the tools to do so (to write and to read). The challenge is no longer efficiency, or pleasure but literacy.

The notion of "information literacy" was introduced in 1974 by the National Commission on Libraries and Information Science[64] and is defined by the American Library Association as "the ability to use printed and written information to function in society, to achieve one's goals, and to develop one's knowledge and potential".[65] Literacies articulate different levels of intellectual processes: not only to recognize the inscriptions, that is to say the signs that one needs to read as forming words and sentences, but also to consider the text as a coherent body of message, and to evaluate how a text makes sense in a document. Literacy eventually includes the understanding of the genre of the document (novel of fiction, manual of use, sacred or political book, etc.) to be able to understand what is the angle of the document: fiction or reality for instance. Literacy also means that people can choose amongst a variety of documents the one they are looking for. It is a "set of abilities requiring individuals to "recognize when information is needed and have the ability to locate, evaluate and use effectively the needed information"(1989).[66] It is an understanding of context and being able to find the right document. Every new media challenges these skills and "new literacies" develop. Media literacies or trans-literacies[67] consist not only in reading and writing but include the ability to recognize and choose relevant media according to task, or to switch from one media to another and follow the diversity of meanings. In HCI, the focus is now on visual literacies and how to assess them.[68]

---

[63] Jeanneret (2007b).

[64] The phrase *information literacy* first appeared in print in a 1974 report by Paul G. Zurkowski, written on behalf of the National Commission on Libraries and Information Science. Zurkowski used the phrase to describe the "techniques and skills" known by the information literate "for utilizing the wide range of information tools as well as primary sources in molding information solutions to their problems". http://en.wikipedia.org/wiki/Information_literacy

[65] http://www.ala.org/advocacy/literacy/adultliteracy

[66] http://infolit.org/about-the-nfil/what-is-the-nfil/

[67] Delamotte Eric, Liquête, Vincent, « Réflexion autour de la notion de 'trans-littéracie informationnelle' »

[68] Boy et al. (2014).

I want to add that this question of literacy is also a political one. Richard Hoggart whose career has covered the fields of sociology, English literature and cultural studies, with emphasis on British popular culture[69] is probably one of the first to give an account of literacy from a political point of view.[70] He belonged to a group of scholars who criticized mass culture as deteriorating popular culture. Media producers know how to embed some part of the popular traditional culture to please their readers. In other words, the media anticipate the way they can be interpreted and how to create a bridge between the reader and the contents. But the readers are always aware of the fact that their culture is so to speak thrown back at them and develop strategies of detachment or irony, in other words strategies of "resistance". The relation to media is therefore a complex one that involves a movement between consumers and producers, readers and writers. When Hoggart speaks of literacy, he has a vision not only of the cultural artifacts, he also includes a history of the different relationships between these media and their audience as well as the way a political conscience is at stake.

From a design perspective, the problem is double: to give users the tools to support their understanding and learning thanks to relevant representations and media, but also to expand the way they are going to communicate and build representations. The first task is not as easy as it seems since the meaning of these representations is not static. It changes in time, space, and culture. The humanities gives designers concepts and tools to be able to assess what things mean in context.[71] The second task in not easy either. New literacies are also a goal for designers: as pointed out by the historian Thierry Bardini, the goal of Douglas Englebart, inventor of the mouse and the hyperlink, was primarily to augment the tools to think with, not to provide with an easy to use device.[72]

The challenge of designing computer systems might, therefore, be more about literacy than ergonomics. The capacity to "read" in context is precisely what new media have to deal with as they offer not only new contents but new ways to shape, access and produce information. In addition to focusing on the teleological efficiency of the tool or the emotional impact of the artifact, designers also have to consider how users can access intellectual contents. In this respect they need to make sure that a common language is being used, that the media offers clues on how to be read, that the tools to write can be used. This issue of new media literacy is at the forefront of political concerns. The different associations that promote knowledge and civic participation, such as UNESCO, emphasize the absolute necessity of training people to read and use these new media to be able to participate in the democracy and be involved in the economic life of their country. In this paradigm, the user is considered more as an "homo sapiens" who agrees to discuss and share world views with others. Designers of Information technologies therefore have a

[69] http://en.wikipedia.org/wiki/Richard_Hoggart

[70] Hoggart (1957).

[71] Harris (1993).

[72] Bardini (2000).

political responsibility as they shape the tools to represent, access, distribute, and use, information. They become the messengers.

3.5 Is the User a "Figure of Speech"?

The previous sections showed that we could have a broader definition of the user that includes the multitasker, the aesthete and the "homo sapiens" who represents and interprets signs. This triple figure of the user works as a typology of properties that designers strive to give to their invention so that they are relevant to their audience. But we could also go deeper in the criticism of the user and her role in design. If the figure of the user plays a role in conception, it has to be more than just a readjustment to the different facets of existing people. Indeed, there is a fundamental difference between observing the uses of today and imagining the uses of tomorrow. Humans and technologies evolve in parallel and new technologies bring out new versions of "humans". There is no causal scheme but an interweaving of society and technology[73,74,,75] The technical object is a series of compromises between actors who carry a social project through their technical propositions. In this respect, scientists who are the first users of their own technology[76] work to embed an *a priori* representation of future uses. Designers create new users because they are focusing on problem spaces that had not been previously considered (Krippendorff, 1989,[77]).

In this section, I therefore want to move beyond user-centered design to open new questions that the rest of the book will continue to examine.

In this chapter, we have encountered many uses of the user:

- The Allegory of the User in research frameworks legitimates the research, gives it its goal.
- The representation of the user that each discipline builds accompany a certain vision of the activity;
- Contrasting user figures triggers a transformation of the disciplines that adjust to a redistribution of the attributes of the user.

These different roles of the "users" in the invention present a very different picture from what is usually expected from user-centered design: that is "real" users interacting with a system that researchers can therefore evaluate.

Throughout this chapter I have used the word "figure" to speak about the user. Is it to say that the user in user-centered design is a total fiction? How can we qualify the relation to reality that the users play in invention? How can we understand how a present user becomes a future user?

---

[73] Akrich (1987).

[74] Akrich (1990).

[75] Akrich et al. (2006).

[76] Vedel (1994).

[77] Krippendorff (1989).

In this last section, I want to be more specific about what a figure means in terms of:

- Representation,
- Transformation.

As we have seen, each discipline claims that it has a proper figure of the user. In other words, they pretend to represent the users in a way that is relevant with what the users are in reality. To put it slightly differently, the different definitions of users allow each discipline to address a specific set of problems. As we have seen in the example, the museum has a definition of the user in relation to its own set of questions and disciplinary concerns: patrimony and conservation, research and education, access and communication. Their model of the user is in some way predictive as they can anticipate and organize their own activity taking into consideration the web of interactions between the different actors of the museum. They define a set of properties of these users that can be met by the institution. We can therefore consider that different definitions of the users are not contradictory but basically answer different kinds of questions. The concept of figure in this instance is not to lessen the quality of user research but to consider what is at stake for the disciplines in terms of representation. Interdisciplinarity from this standpoint is a way to add the different definitions to have a richer perspective on the situation.

However, the role of the figure to represent a model of the actors and activities is challenged by the invention. While the model, either qualitative or quantitative, is developed on the basis of observations, the simulation of what could be a user of a new artifact, needs to confront and deregulate the set of properties of the representative users.

What the experience with PLUG showed was that each definition of the user brought by each discipline could not totally encompass the complexity of a real person. In fact, even put together, the disciplines will not get the full picture. However, the different definitions can change the definition of the user within each discipline. This is where the actual innovation takes place. The figure of the user serves as a trigger to unsettle the scientific disciplines, make them recognize that they do not have a full picture and therefore to seek new definitions. At the beginning of the invention, contrasting different figures of the user triggers the design of a new ensemble of parameters. Conception is therefore based on a destabilization of the concept of user for each discipline. At the end of the process, the inventors converge on a new set of figures of the user through the new artifact. User studies then help adjust the invention to situations in the field. At first, the pluridisciplinarity of design is therefore less a convergence on one single definition, or the organization of different points of view on the same question, but more the destabilization of disciplines that forces them to question their assumptions and reconsider their concepts.

The concept of figure is here useful to understand the transformations that take place. The semiotician Louis Hebert, elaborating on the group μ, Zilberberg, and Rastier, considers how a structure is made of terms, relations, and operations. What I am interested here are the operations.

An operation is a process, an action by which an operating subject characterizes or transforms an object (whether the object is a relation, a term, or an operation).

- Operations of characterization isolate properties of an object by decomposition (mental), classification, typing or categorization, comparison, and other processes.
- Operations of transformation (1) produce objects (through creation ex nihilo, emanation from a type, or construction from materials known to be pre-existing), (2) destroy them (through annihilation, i.e. with no remaining trace, or through complete deconstruction) or (3) transform them".

While the figure of the user within the discipline is an operation of characterization, the figure of the user in invention is an operation of transformation. Both are related in so far as the transformation can take place thanks to several types of operation that start from the characterization (and probably the fact that it decomposes the attributes of the "term"). Contrasting different characterizations allows us to come up with a new figure of the user.

Hebert counts six operations of extensity:

- (1) addition (or blending: e.g., A becomes A, B);
- (2) deletion (or sorting: e.g., A, B becomes A);
- (3) substitution (e.g., A, B becomes A, C);
- (4) permutation (e.g., A, B becomes B, A);
- (5) displacement (e.g., an eye displaced onto the belly of a monster) and
- (6) continuance (e.g., A, B remains A, B).

The figure is a way to play with the different parameters of different users definitions. Entering this plane of representation opens up a flexible plane of reconfiguration. From a design perspective, it is therefore critical to understand how a narrative of future uses, and future users, is created by engineers and designers and embedded in the new artifacts. My hypothesis is that conception is based on "science fictions" that offer a vision of the future. In this respect, the role of social scientists is to be reconsidered. Indeed, their contribution is not, as it would seem, to provide hard social data that directly feed in the radical invention. Instead, they provide narrative material that feeds the creation of scenarios, imaginary stories that are indirectly related to what our society is today. As will be developed in Chap. 4, we need to look at the poetic practices of inventors who reintroduce the user indirectly.

References

Aarts, E., & Appelo, L. (1999). Ambient Intelligence: thuisomgevingen van de toekomst. *IT Monitor*. 9/99, 7–11, The Netherlands.

Akrich, M. (1987). Comment décrire les objets techniques? *Techniques & Culture*, 49–64. https://halshs.archives-ouvertes.fr/halshs-00005830/.

Akrich, M. (1990). De la sociologie des techniques à une sociologie des usages. *Techniques & Culture*, 83–110. https://halshs.archives-ouvertes.fr/halshs-00081710/.

Akrich, M., Callon, M., & Latour, B. (2006). *Sociologie de la traduction : Textes fondateurs*. Paris: Presses de l'Ecole des Mines.

Bannon, L. (2011). Reimagining HCI toward a more human-centered perspective. *Interactions, July-August*, 50–58.

Bardini, T. (2000). *Bootstrapping: Douglas Engelbart, coevolution, and the origins of personal computing*. Stanford: Stanford University Press.
Becker, H. S. (1966). *Outsiders, studies in the sociology of deviance* (reprintth ed.). London: The Free Press.
Berthelot, J.-M. (2001). *Epistémologie des sciences sociales*. Paris: Presses Universitaires de France – PUF.
Blum, A. (2013). *Tubes: Behind the scenes at the internet*. London: Penguin.
Boehner, K., Sengers, P., & Warner, S. (2008). Interfaces with the ineffable: Meeting aesthetic experience on its own terms. *ACM Transactions on Computer-Human Interaction (TOCHI), 15*(3), 12.
Boy, J., et al. (2014). A principled way of assessing visualization literacy. *IEEE Transactions on Visualization and Computer Graphics, 20*(12), 1963–1972.
Bucciarelli, L. L. (1996). *Designing Engineers*. Cambridge, MA: The MIT Press.
Castells, M., Cardoso, G., Paul, H., & Nitze School of Advanced International Studies Center for Transatlantic Relations. (2006). *The network society: From knowledge to policy*. Washington, DC: Center for Transatlantic Relations, Paul H. Nitze School of Advanced International Studies, Johns Hopkins University.
Crilly, N. (2010). The roles that artefacts play: Technical, social and aesthetic functions. *Design Studies, 31*(4), 311–344. doi:https://doi.org/10.1016/j.destud.2010.04.002.
Cummings, J. N. (2005), octobre 1. Collaborative research across disciplinary and organizational boundaries. *Social Studies of Science, 35*(5), 703–722. doi:https://doi.org/10.1177/0306312705055535.
Davallon, J. (2000). *L'exposition à l'oeuvre*. Paris/Montréal: Editions L'Harmattan.
Davidson, D., Kelley, H. & Kücklich, J. 2007. *Space time play* (F. von Borries, S. P. Walz & M. Böttger, 1st ed.). Boston: Birkhäuser Architecture.
de Ruyter, B. (Ed.). (2003). *365 days' ambient intelligence research in HomeLab*. Eindhoven: Philips Research.
Delamotte, E., & Liquête, V. (2010). La trans-littératie informationnelle. Eléments de réflexion autour de la notion de compétence infocommunicationnelle scolaire et privée des jeunes. *Recherches en Communication, 33*, 17–34.
Ducatel, K., & Bélgica. (2001). *Dirección General de la Sociedad de la Información Comisión Europea (Bruselas, Scenarios for ambient intelligence in 2010)*. Luxembourg: Office for official publications of the European Communities.
Durkheim, E. (1997). *Suicide: A study in sociology* (reissue edition). Glencoe, III.: Free Press.
Gentès, A., & Jutant, C. (2009). Pervasive gaming: Testing future context aware applications. *Communications and Strategies, 73*, 81–104.
Halskov, K., & Hansen, N. B. (2015), février. The diversity of participatory design research practice at PDC 2002–2012. *International Journal of Human-Computer Studies, 74*, 81–92. doi:https://doi.org/10.1016/j.ijhcs.2014.09.003.
Harris, R. (1993). *La sémiologie de l'écriture (CNRS langage)*. Paris: CNRS editions.
Hoggart, R. (1957). *The uses of literacy: Aspects of working-class life*. London: Penguin Classics.
Hsi, S. (2002). *The Electronic Guidebook: A Study of User Experiences using Mobile Web Content in a Museum*. International Workshop on Wireless and Mobile technologies in Education. Danvers: IEEE.
Ilpo Koskinen, K. B., & Mattelmaki, T. (Eds.). (2003). *Empathic design: User experience in product design*. Helsinki: IT Press, Edita Prima Ltd.
Jeanneret, Y. (2007a). Usages de l'usage, figures de la médiatisation. *Communication et langages, 151*(1), 3–19. https://doi.org/10.3406/colan.2007.4629.
Jeanneret, Y. (2007b). *Y a-t-il (vraiment) des technologies de l'information ?* Villeneuved'Ascq: Presses Universitaires du Septentrion.
Jordan, P. (2000). *Designing pleasurable products: An introduction to the new human factors*. London: Taylor & Francis.

Katz, E., Blumler, J. G., & Gurevitch, M. (1973). Uses and gratifications research. *The Public Opinion Quarterly, 37*(4), 509–523.
Keinonen, T. (2010). Protect and appreciate–notes on the justification of user-centered design. *International Journal of Design, 4*(1), 17–27.
Krippendorff, K. (1989). On the essential contexts of artifacts or on the proposition that "design is making sense (of things)". *Design Issues, 5*(2, Spring), 9–39.
Latour, B. (1983). Give me a laboratory and I will raise the world. In *Science observed* (pp. 141–170). London/Beverly Hills: Sage.
Le Marec, J. (2012). *Publics et musées : La confiance éprouvée*. Paris: Harmattan.
Lévy, P. (1994). *L'Intelligence collective : Pour une anthropologie du cyberspace*. Paris: La Découverte.
Mackay, W., et Fayard A.-L. (1997). HCI, natural science and design: A framework for triangulation across disciplines. In *Proceedings of the 2nd conference on Designing interactive systems: Processes, practices, methods, and techniques* (pp. 223–234).
Manovich, L. (2002). *The language of new media*. Cambridge: The MIT Press.
Maslow, A. H. (1943). A theory of human motivation. *Psychological Review, 50*(4), 370–396.
McCarthy, J., & Wright, P. (2007). *Technology as experience*. Cambridge: The MIT Press.
Montola, M. (2005). Exploring the edge of the magic circle: Defining pervasive games. In *Proceedings of DAC*, p. 103. http://remotedevice.net/main/cmap/exploringtheedge.pdf
Nigel Cross & Design Research Society (1972). *Design Participation: Proceedings of the Design Research Society's Conference, Manchester, September 1971*. Academy Editions.
Norman, D. A., & Draper, S. W. (Ed.). 1986). *User centered system design: New perspectives on human-computer interaction* (1st ed.). Boca Raton: CRC Press.
Prahalad, C.K., & Ramaswamy, V. (2004), janvier. Co-creation experiences: The next practice in value creation, *Journal of Interactive Marketing*, *18*(3), 5–14. doi:https://doi.org/10.1002/dir.20015.
Redström, J. (2008). RE:Definitions of use. *Design Studies*, *29*(4, juillet), 410–423. doi:https://doi.org/10.1016/j.destud.2008.05.001.
Ritzer, & Jurgenson, N. (2010). Production, consumption, Prosumption the nature of capitalism in the age of the digital 'prosumer. *Journal of Consumer Culture, 10*(1), 13–36.
Rochlitz, R. (1998). *L'art au banc d'essai: esthétique et critique*. Paris: Gallimard.
Sanders, E. B.-N. (2002). From user-centered to participatory design approaches. In J. Frascara (Ed.), *Design and the social sciences: Making connections* (pp. 1–8). London: Taylor & Francis.
Satyanarayanan, M. (2001). Pervasive computing: Vision and challenges. *Personal Communications, IEEE, 8*(4), 10–17.
Scott, P. (1991). Levers and counterweights: A laboratory that failed to raise the world. *Social Studies of Science*, *21*(1, janvier 2), 7–35. doi:https://doi.org/10.1177/030631291021001003.
Simatic, M., et al. (2009). "Plug: Secrets of the museum": A pervasive game taking place in a museum. In *Entertainment computing–ICEC 2009* (pp. 302–313). Springer. http://link.springer.com/chapter/10.1007/978-3-642-04052-8_44.
Souchier, E., Jeanneret, Y., & Le Marec, J. (2003). *Lire, écrire, récrire: Objets, signes et pratiques des médias informatisés*. Paris: Bibliothèque Publique d'Information.
Vedel, T. (1994). *Introduction à une socio-politique des usages, in André Vitalis «Médias et nouvelles technologies »*. Rennes: éditions Apogée.
Virilio, P. (1986). *Speed & politics*. New York: Semiotext(e).
Von Hippel, E. (2006). *Democratizing innovation*. Cambridge, MA: The MIT Press.
Xie, C., Bagozzi, R. P., & Troye, S. V. (2007). Trying to prosume: Toward a theory of consumers as co-creators of value. *Journal of the Academy of Marketing Science*, *36*(1, août 1), 109–122. doi:https://doi.org/10.1007/s11747-007-0060-2.
Zelkha, E. (1998). *The Future of Information Appliances and Consumer Devices, Palo Alto Ventures, Palo Alto, California* (unpublished document).
Zwick, D., Bonsu, S. K., & Darmody, A. (2008). Putting consumers to work co-creationand new marketing govern-mentality. *Journal of Consumer Culture, 8*(2), 163–196.

Chapter 4
The Poetics of Invention

4.1 Questions and Methods

Why call a technology Bluetooth after a king of Denmark? Why choose a sheep to represent a research project? And why tell all kinds of stories when one should concentrate on serious matters directly related to engineering sciences like algorithms or programming? We could dismiss the naming and storytelling activities on the grounds that they are commercial, futile, or even whimsical. They do not seem to bring anything to the scientific validation, or promotion, of a research project. However, it is hard to overlook the fact that naming a technology, creating logos, and telling stories, are actually part of any engineering project and therefore the question remains: what is their role and value in research practices? Looking at examples ranging from the naming of a field such as "artificial intelligence" to the christening of a project on distributed mobile networks like "SAFARI", I make the hypothesis that research and engineering could be analyzed as poetic activities. Beyond this, I also contend that naming, telling, putting into images, are not only the final touch of a marketing program but part of the expansive part of the design process. If this hypothesis is true, ignoring these practices could be detrimental to strategies of innovation. In this chapter, I therefore look at the place and role of symbolic presentations as part of the making of invention and design. Here, elaborating on the philosopher of science and technology Simondon who defines invention as articulating imagination to cultural and social context through a tangible artifact,[1] I further argue that the tangible artifact is not the only production that defines the invention. The latter is made of poetic and narrative creations that eventually generate the tangible artifact. Logical demonstrations or negotiations between actors are not enough to support an invention that, in fact, benefits from subjective and cultural associations, **and** poetic activities related to humanistic and literary practices. Researchers and designers use and work on symbolic representations,

[1] Simondon and Chateau (2005).

A. Gentes, *The In-Discipline of Design*, Design Research Foundations,
DOI 10.1007/978-3-319-65984-8_4

poetical concepts and narratives that expand the observations and definitions that they obtain within their methodological framework. This chapter addresses poetical practices within research and design, and shows how engineering sciences articulate with imagination. My claim is that poetical practices that focus on the subjective and cultural associations need to be considered as an essential part of technical research, in particular because they organize a writing of the future. To study them, I need, first, to define what poetics are in contrast to rhetoric and demonstration. As there is a large body of literature on these discursive and literary questions, my purpose is not to sum them up. I essentially point out certain aspects that I find relevant to the understanding of conception.

I studied several poetical practices: how researchers in engineering and design name new technologies so that they become "common names"; how researchers christen their technical projects with "proper names" like family names; how they use logos to perform the ongoing research; how they tell stories about the invention to be. Whenever possible, I tried to see the connections between the invention process as it took place and the poetic productions. The method is therefore based on ethnographic participation (the pragmatic dimension of this research) and a collection of productions (the semiotic dimension of this research). I also want to briefly suggest that identifying, analyzing and using a corpus, which is a fundamental method of the humanities, is necessary to the understanding of methods of conception.

Why and how do we use corpuses? Semiotics scholars define a corpus as a group of texts (linguistic and/or visual) that belong to the same genre.[2] Texts of the same genre share structural and pragmatic characteristics in so far as they are not only similar in style, narrative structure, or visual organization, but also as they are discussed by social actors as targeting the same audiences, for the same purposes, with similar textual strategies. A corpus therefore is based on formal properties that are generally considered similar by a group of actors. For instance, science fiction authors, editors, critics, and readers can agree that certain novels belong to this group of literature based on formal and experiential qualities. Of course, some novels can be introduced or removed from the genre, when the consensus is that they either add to the genre (and even make it evolve), or on the contrary no longer correspond to the definition of the genre and its proper attributes. Beyond the formal properties of the examples here, one must therefore understand that they are generally considered as belonging to the same group of intellectual productions by the actors. Nevertheless, the constitution of the corpus does not pretend to be exhaustive. It is always but a segment of the actual production. The purpose is to gather enough material to answer a research question. The method therefore consists in defining a corpus of textual elements and considering what are the shared elements and their purpose in the text and in the experience of their readers, spectators, and

[2] Rastier (2001).

users. To do so, the humanities researcher can and will base her selection following the cue given by the actors. But she can also add new elements and compare productions with the hypothesis that they offer similar aesthetic or social experiences. For my research, I selected examples over a diversity of situations: condensed forms of identity through names or logo and extended forms of identity through staging the invention within narratives. My purpose is to emphasize the similarities over different examples rather than to focus on the differences that should be further explored. In other words, I have gathered documents that belong to the same genre because they are produced for the same type of readers, on the same occasions, by the same type of actors. Then, I analyze from a literary perspective what their formal and pragmatic traits are so as to distinguish nuances and give a better definition of the experiential qualities of these creations in particular to answer my research question, namely what are their expansive and performative proprieties.

4.2 Poetic Versus Rhetoric

We may think of science and technology as fields that are safe from the hassles of society and the fantasy of imagination, but this chapter contends that the process of invention depends on dealing with the social space and creating words or images that perform the invention. In Chap. 5, I will deal with the social and, more precisely, the debatable space of invention. Here, I want to focus on what it means to create words and images that actually project the invention.

The scientific discourse is based on logical demonstrations that mark it from common knowledge and language.[3] But making science is also about strategies to defend a territory, to plead for a cause, to fight enemies, to gather followers. Actors build a full-fledged rhetorical strategy to support their project. This part of science has been thoroughly analyzed by sociologists of sciences.[4] I will briefly evoke some of their findings. Mostly they argue that texts (including images like drawings, photographs and videos) that are produced in the course of a research project pursue extensive and defensive functions. However, I think that they also contribute to the invention because they project an image of the technology. In other words, they are part of a specific language game that speaks "about" the invention, and also makes it happen. This language game not only describes or defends research findings, it is also performative and expansive.

[3] Chalmers (1999).

[4] Hackett and Society for Social Studies of Science (2007).

4.2.1 *From the Rhetoric of Science...*

Sociologists of science build their analysis of how science is made on the basis of a keen understanding of language games played by the actors through different channels of communication. They pay attention to the rhetorical skills that agents deploy and that can win their case. I find it useful at this point to go back to the definition of rhetoric, because since the origins of the concept, it is contrasted with other forms of speech that have organized the way we think of different worlds: the everyday world of passions, the realm of science and truth, and the space of our imagination.

Rhetoric was described by Aristotle as an art of persuasion that every citizen should learn in order to debate about things in the city, and, eventually, to be able to defend himself in court.[5] Two points are central to his demonstration: in a democratic society, things are being discussed and judged and while scientific and rational arguments are important, winning the audience often means that a demonstration should be discarded in favor of more appealing presentations. Aristotle, in a very practical manner, considers that persuasion prevails when education and instruction cannot take place. It is not always possible to teach, to explain, to train, therefore one must convince by other means.

> Before some audiences not even the possession of the exactest knowledge will make it easy for what we say to produce conviction. For argument based on knowledge implies instruction, and there are people whom one cannot instruct. Here, then, we must use, as our modes of persuasion and argument, notions possessed by everybody.[6]

The goal of rhetoric is for people to fulfill their social duty and to keep their power and place in society. The whole point of rhetoric is therefore to reach out to an audience and to convince it. Primarily, men as political animals have to decide what is fair or not, what is probable or not, and this kind of decision is part of a social process of communication. To do so, several types of arguments can be used that are either centered on the orator's privileged relationship to the subject, or on the particulars of the subject itself, or eventually on the foreseen relationship of the audience with the subject. From Aristotle's point of view, rhetoric is a specific speech strategy that is useful not only because people do not have the "exactest" knowledge about things, but because the realm of justice is one that involves values and ethics that will not be reduced to a scientific demonstration. In other words, rhetoric is pitted against discourses of truth but also against imaginary discourses and fictions.

The sociologist of sciences and technologies – Bruno Latour's – contribution,[7] has been to analyze the form and role of rhetoric in the scientific world. He has showed that making science, implies defending it and therefore elaborating a rhetoric that convinces the different stakeholders. For instance, he points out that scientific papers

---

[5] Aristotle and Lam (2014).

[6] Ibid. p. 6.

[7] Latour (1988).

are built around a defensive strategy that finds allies (other scientific authors and papers) as a way to prevent criticism. They "enroll" other papers that have been written to prevent contradiction. Latour speaks of "stratification"[8] that is to say the building of a system of authority within the paper. This system also excludes those researches that do not follow the same direction. Research and scientific activities are therefore like a court where participants build up the defense of their cause. For Latour, this winning rhetorical strategy makes it difficult for scientists to produce certain types of discourses. In particular, he points out that popularization is difficult because scientific texts are written to exclude people not to include them. What is more, scientific controversies need strong uncompromising positions that do not leave any space for dreams or imagination because they need to be grounded in "truth". To rapidly sum up his contribution, he has effectively shown that, as far as science is concerned, we need to consider rhetoric as part and parcel of the discursive strategies of scientists. This "relativism" of science has been either praised or heavily criticized as it seems to undermine the epistemological claims of the disciplines. For me, the opposition does not tackle a fundamental issue that is neither relativist nor strictly methodological but which looks at how, in science, as in other fields of intellectual productions, something **new** emerges. Asking this question, I place design at the center of my reflection but also engineering sciences as they explicitly look for new technical dispositives. Without criticizing the sociological dimensions of sciences nor disparaging the sciences' claim for truth, I want to understand how sciences can be properly inventive. To do so, we need to look closely at the way sciences actually include heterogeneous discursive elements.

4.2.2 ... To the Poetics of Science

The strategic vision of scientific prose is determinant but it is only a part of the picture. Repeatedly, I came across productions that did not fit the "defensive" pattern. Calling a research program "Popeye" can hardly count as a serious scientific expression nor as a particularly fierce defense of the research in question... What is at stake in images and stories that use more fictional metaphors and visualizations than hard core arguments? How can we account for these scientific activities that look more like humorous, poetic endeavors with, at their heart, popularization processes?

While over the centuries, poetics has been reduced to a collection of stylistic traits, it is in fact an art of production that pursues the perfection of a designed object. Aristotle's introduction to poetics[9] in the book that he devotes to art (more precisely, theater, poetry, and painting) can give us useful guidelines in distinguishing the specificity of this type of production. Without offering an extensive account of the philosopher's work, I want to point out two elements that can be useful to

[8] Latour, p. 120–127.

[9] Aristotle (2005).

understand design and engineering: first the properties of poetics, second the type of work that is involved. For the philosopher, the purpose of poetry is not to use language as a persuasive means to create a social consensus but to create a vision of the world and humanity. Artists transcend the here and now, because they unfold either **latent** or **possible** aspects of society. The artist sees through what is already here to consider what could be. The first goal of these productions is not to teach, defend, nor demonstrate like rhetorical or dialectical discourses, but to open new alternatives that are relevant to our experience. Hence a double evaluation: first, they are evaluated on their capacity to augment the number of possible experiences; second they are evaluated according to their relevance regarding our experience of the world. A poetic work does not describe scientifically but it still includes an indirect relation to its contemporary world. In this respect, according to Aristotle, poetry has more to do with philosophy because it creates ways of looking at the world. This difference between science on the one hand and poetry and philosophy on the other hand, is further developed when Aristotle stresses the difference between the poet and the historian. This crucial difference is probably at the root of this book on design and its relationship to social sciences and humanities.

> The poet and the historian differ not in that one writes in meter and the other not; for one could put the writings of Herodotus into verse and they would be nonetheless history, with or without meter. The difference resides in this: the one speaks of what has happened, and the other of what **might be**. Accordingly, poetry is more philosophical and more momentous than history. The poet speaks more of the universal, while the historian speaks of particulars. It is universal that when certain things turn out a certain way someone will in all likelihood or of necessity act or speak in a certain way—which is what the poet, though attaching particular names to the situation, strives for (*Poet.* 1451a38–1451b10).

The historian and the artist create a discourse but one translates the events (observations, samples, etc.) into an explanatory discourse while the other introduces some "play" in our relationship to the world to change both the way we look at the world and what it is made of. For me, this distinction is fundamental to understanding how design and engineering research, or changes in scientific paradigms, impact the world.

The second question concentrates on how to achieve these goals. Aristotle uses the word *Dunamis* or *dynamis* that means possibility or capability. The design work consists of exploring a variety of productive possibility and capability while processing, working on and with a variety of materials: either language (styles, genres,[10] or pigment or sound[11]). Playing with words, inventing stories, making images are therefore intensely poetic activities. Within a medium (whether words, wood, or sounds, or anything else) the artist finds the best way to make perceptible this potential world of experience.

More recently, poetic studies have focused on the creative process itself and its specificity.[12] In particular, the linguist Jakobson proposed a model of language that

---

[10] Genette (2014).

[11] Gentes (2012).

[12] Preminger et al. (1993).

looks at how the poetic function forces readers or listeners, more than other linguistic functions, to attend to the signifiers in linguistic signs, and not so much to the signified. The poetic function of language focuses on its own material. It is opposed to the referential function of language that connects words to things[13] so as to allow effective communication. Referential and poetic functions of language are further analyzed by semioticians, who distinguish denotation and connotation. Denotation is about literal meaning and is essentially descriptive while connotation covers subjective and cultural associations.[14] While Jakobson's treaty is about language, the history of visual arts and in particular painting can be seen as a gradual emancipation from the demands of the denotative, and as a gradual claim for a specific place for the poetic space that it represents. The French writer, poet, and literary analyst, Paul Valéry also described this poetic stance as a suspension of the use of language as a persuasive means to create a social consensus, and as an endeavor to transcend the here and then, because the writer or artist unfolds either latent or possible aspects of the world.[15] For Meschonnic too, a specialist of poetry and translation, the creator questions what could be.[16]The artist taps into and reveals the potential of a medium to transform and interpret the world. Her composition can be called expansive, not because it foretells the future but because it produces new meaning. Playing with words, inventing stories, making images are not only there to teach, or to defend, or to demonstrate. They expand the way we think about the world.

Poetic practices such as imaging and visualization within research projects have recently been the subject of social studies of scientific texts,[17] in ethnography for instance.[18] Scientific images have been noted for their aesthetic ambition.[19] Sociologists of science, Burri and Dumit point out the "engagement" that images produce and in particular how they are made instrumental in the production of scientific knowledge.[20] While Latour, Lynch and Woolgar[21] stress how images tend to objectify knowledge and to produce yet another argument in the rhetoric of science, we follow those who note the fecundity of images with respect to creativity and invention. Images in this respect are considered epistemic creations.[22] Elaborating on this literature, I find that the "visual turn" (as qualified by Mitchell[23]) is only part of a broader set of poetic practices that not only define epistemic communities, support argumentation, or contributes to knowledge, but also are part of the conception of new objects.

---

[13] Jakobson (1981).

[14] Barthes (1973).

[15] Valéry (1957)

[16] Meschonnic (1970).

[17] Olohan (2000).

[18] Clifford et al. (1986).

[19] Sicard (1995).

[20] Burri and Dumit (2007).

[21] Lynch and Woolgar (1990).

[22] Cambrosio et al. (1993).

[23] Mitchell (2002).

In my experience, the poetical process participates in the definition and dynamics of research projects. By poetical process I do not mean the scientific texts as such or the technological achievements, but all the productions – video, text, photo – that strive to carry the meanings of the technology, and in doing so influence the way researchers invent their projects and perform their ideas. Several aspects of this poetic process are exposed in this chapter: the part played by naming, the importance of narratives, the role of different media in opening the meaning of the technology.

4.3 Speaking of Which… Word Invention

In a field where words and expressions are invented every day, people can use the same words with obviously very different meanings. These "suitcase" words are not necessarily as dense as "consciousness" or "emotion" studied by Marvin Minski,[24] but they still cover a lot of ground. For example, in the ADAM project on peer to peer and distributed architectures, the words "centralized" or "decentralized" and "hierarchical" and "non-hierarchical", could convey a very technical meaning in terms of routing protocols but also a political meaning in terms of who does what and how information circulates. People from social sciences focus on applications and power struggles while network engineers might think more about the efficiency and robustness of the infrastructure. But while it is important to discuss these concepts and collectively agree on their meaning (until they are redefined again) I find that the fuzziness of the words has some advantage in a group. The absence of a unique definition not only helps with everyday interaction, it also opens a possibility to play on the polysemy of the words so that everybody can eventually change the meaning of these words within their own discipline. Rather than communicating a specific meaning, participants can implicitly agree on the limits of their knowledge and enrich their own vocabulary with words that they leave open for interpretation.

As the expert on languages David Crystal suggests: a "traditional account of the history of languages tends to minimize or ignore the inherent messiness found in real-life linguistic situations".[25] This chapter wants to show how messy the whole linguistic practice of words invented for new information and communication technologies really can be but also how it drives invention. In the following section, I will present different stylistic choices: scientific puns, acronyms, logos and look at their role in the creative process. As each program included researchers in design and in engineering, I have tried to sum up how the different participants pragmatically defined the properties of their stylistic choices in the design process.

[24] Minsky (2007).

[25] Crystal (2004). p. 20.

4.3.1 Scientific Puns: Play on Words and Definitions

Studying designing engineers, Bucciarelli made two main observations. First, designers have "object world" languages that allow them to be specific on their area of expertise but that they also need to partly share with people from other disciplines. His hypothesis is that design is a collective process consisting in understanding part of these object world languages. Second, designers invent words[26]: "Naming is designing".

> The invention of a name for a part of the design, for a piece of the action, is designing. A name, e.g. 'archiving of images', once established, while it may conjure up different visions of form and function within the minds of different design participants, serves to label a particular focus of common concern. It defines the ballpark so to speak, providing an arena for design moves within the context of the subculture's traditions, last year's model, the experience and competencies of participants.[27]

Now, I want to look a little closer at what kind of names are given, when and for what creative purpose.

Technologies that are invented need to be given a name. The flourishing invention of words in the information and communication sector is fascinating not only because it covers a large number of technical innovations in that field but also because it displays a rich diversity of poetic strategies. Some words describe the technology: "802.11"[28]; some are the result of brand work, "Wi-Fi" that supposedly stands for "wireless fidelity"; some rely on metaphors: "ad hoc" networks; others completely drop out the technical description and look like an arbitrary choice as the communication standard "Bluetooth". I want to go back to some of these names, starting with some of the less explicit ones.

The term "Bluetooth" for instance seems in no way related to what it qualifies. A leading actor of the "Bluetooth" norm, Jim Kardach, posted on his blog that the diversity of descriptive technical names was such that participants in the normalization meetings were getting frustrated with the aridity and complexity of using a denotative appellation for this new type of short-range radio frequency.

> *It became apparent the need to have a single name; as Intel would talk to people about "Biz-RF", Ericsson about "MC-Link" and Nokia about "Low Power-RF", which also created confusion...* (Kardash 2008)

He says that the name King Harold Bluetooth of Denmark, who unified the various Danish regions in the tenth century, emerged in discussions and post conferences between researchers who were attending normalization meetings because of a book that one of them (Sven Mattisson) was reading at the time: *The Longships* by Frans G. Bengtsson, as well as another book that Kardash was reading about Danish history: *The Vikings* by Gwyn Jones. These readings were probably influenced by the culture of some of the participants[29]. While it may have started as a joke and a

---

[26] Bucciarelli (2002).

[27] Bucciarelli (1988).

[28] "the first wireless networking standard", https://en.wikipedia.org/wiki/IEEE_802.11

[29] http://www.eetimes.eu/scandinavia/206902019?cid=RSSfeed_eetimesEU_scandinavia

code name between people "in the know" and for lack of anything better, it finally stuck and became the official name of the standard[30]! Rather than keeping complicated names or acronyms that were far too descriptive and technical, the participants preferred the name Bluetooth.

The story of the name Wi-Fi is another example of a strategy departing from the denotative expression: IEEE 802.11b. It was thought as a brand, misinterpreted as a description, and finally triggered an incredible amount of graphical and poetical interpretations and practices. However, at the time of its deployment between 1999 and beginning of 2000, a number of journalists pointed out that: "Wi-Fi doesn't stand for anything. It is not an acronym. There is no meaning".[31] Obviously, this statement needs to be mitigated. A brand means something.[32] In an article, quoting Phil Belanger, a founding member of the Wi-Fi Alliance who presided over the selection of the name "Wi-Fi", we learn that:

> Wi-Fi and the ying yang style logo were invented by Interbrand. We (the founding members of the Wireless Ethernet Compatibility Alliance, now called the Wi-Fi Alliance) hired Interbrand to come up with the name and logo that we could use for our interoperability seal and marketing efforts. We needed something that was a little catchier than "IEEE 802.11b Direct Sequence. […]
>
> The only reason that you hear anything about "Wireless Fidelity" is some of my colleagues in the group were afraid. They didn't understand branding or marketing. They could not imagine using the name "Wi-Fi" without having some sort of literal explanation. So we compromised and agreed to include the tag line: "The Standard for Wireless Fidelity" along with the name. This was a mistake and only served to confuse people and dilute the brand.[33]

This testimony is interesting as it shows a controversy between tenants of a descriptive, denotative name with a literal description and tenants of a name that would be "catchier" and more suggestive to invent new uses.

In our work together, Ted Selker described the naming process of a digital device that he invented at IBM: the Trackpoint device that earned him the title of IBM fellow.[34] The project was first called the "Jstick" because people could use their right index finger on the j key as a joystick. For ergonomic and technical reasons, Ted explained that the device was later placed between the "g" and "h", and got christened with a functional name: "pointing stick" in press releases and presented at the conference: Interact in 1990.[35] But various people proposed that if a 6 inch handle that users wiggled was named a joystick, this little contraption could be named "joy nub", especially after an IBM's design consultant, Richard Sapper, chose red for the device. Already for this tiny thing, four strategies were apparent: either the name followed the obvious position on the keyboard. It relied on a topological description

---

[30] Read more at http://www.todayifoundout.com/index.php/2011/10/the-bluetooth-standard-is-named-after-a-10th-century-scandinavian-king/#JWwfr6cWLWgD1XrS.99

[31] http://community.arubanetworks.com/t5/Technology-Blog/What-does-Wi-Fi-stand-for/ba-p/256914 (retrieved December 2016).

[32] http://interbrand.com/work/catchier-than-ieee-802-11b-welcome-wi-fi/

[33] http://boingboing.net/2005/11/08/wifi-isnt-short-for.html

[34] Gentes and Selker (2013).

[35] Rutledge and Selker (1990).

of the device: "Jstick"; either the underlying technical system provided with a name: hence the choice of "pointing stick". But conversations also focused on a descriptive "subjective" name and people described the appearance and the gesture and considered it as obviously erotic: the expression "joynub" came up. Finally, the company chose a name that came from a line of products: "TrackPoint". The name relied on the activity of the user: tracking down a point or pointing at the screen, and concentrating on the effect on the screen.

On the other side of the technological spectrum, are the birth of entire fields of research. Here, too, finding a name for the research is not only a marketing ploy, it engages the expansive nature of the invention. Ted Selker also pointed to the case of "Artifical intelligence". John McCarthy explains how he chose to use the name: "Artificial Intelligence" in the 1950s[36] to name the field that he was helping to found. For him, Artificial Intelligence (AI) would captivate people's imagination more than the other term that had been considered and used: "Machine Intelligence". In addition to his love of science fiction, he supposed that people would be more interested in exploring human intelligence as well as more capable machines and he wanted to connect the two. Instead of simply describing the work as making yet another engineering field, the expression Artificial Intelligence showed the ambition of a field that wanted to invent technologies that would equal the workings of the human brain.

Whether or not all these names were given for marketing reasons (or in any event to make sure that they supported a positive image for the public) is beside the point. Even when it was the case, I find it is more important to consider whether this process of naming somehow helped the invention to be and how it did so.

If we go back to naming strategies, several constraints must be recognized. Naming a device, a technology, is obviously part of the global process of invention. But the name of an invention has to be formed so that it will become a "common name", that is to say a designation that will apply to a class of objects. While the process can start as a branding activity (like Wi-Fi) the name will not be considered as a proper brand (with copyright) but as a standard appellation for everyone to use.

Then, if we focus on the names themselves, different strategies appear:

- the strictly technical definition, made of a few words or of a number (like the radio wave frequency: 802.11) is what Susan Leigh Star and James Griesemer call a "lowest common denominator".[37] From an engineering perspective, these "names" can synthesize the issue: sufficiently specific so that there is an agreement on what the technology stands for, sufficiently abstract to encompass different stakes, issues, and even opposite views of what the technology can bring to the user. The alternatives to Bluetooth are good examples: *"MC-Link"*, *"Low Power-RF"*. These are names that do not engage with a social practice, cultural contents, or subjective benefits and values. They play a game of neutrality, which

[36] McCarthy et al. (1955).

[37] Star and Griesemer (1989).

is often the claim of technologists. This naming strategy is therefore both synthetic and prudent.

- less abstract but also closer to a description, is the Trackpoint that Ted Selker invented. His story was an interesting one in that it qualified a small element in a bigger technology. The final name describes the user's activity. Such a naming strategy bridges the gap between technology and users which is no surprise in this particular instance since the inventor is in the field of Human Computer Interaction and has a strong interest in the social impact of technologies.

What these first two denotative types show is that giving a name based on a description is less easy than it would seem. The name either points to the underlying technology or to its surface properties, or designates how it works or how it is perceived from a user perspective. Such choices open the process of interpretation in different directions. This is, of course, also true with more connotative names.

- The metaphors like "artificial intelligence" or even the invented name "Wi-Fi" offer more leeway and depth for interpretation. First, these names support a "condensation" process – to use a term that Freud used to describe the work of imagination in dreams.[38] A "Suitcase" word (a signifier) in fact relates a variety of signified objects. Second, they work as oxymorons bringing together different worlds and changing how we look at the universes they come from. The expression "Artificial Intelligence" changes both the way we think about the artificial and about intelligence. The expression Wi-Fi brings together a technical syntagm: "Wireless", while "Fidelity" introduces a value.

Metaphoric names, like "Artificial Intelligence" or Wi-Fi, though less accurate than the previous denotative examples, deepen the relationship of the technology with other products and uses because they evoke linguistic and social codes, rhetorical and ideological conventions and therefore enlarge the cultural references of the innovation. From a conceptive perspective, these expressions allow for more freedom of associations. As a matter of fact, McCarthy's goal was to stimulate the imagination and the research thanks to the evocative metaphor of intelligence and the open ended nature of the word artificial that no longer related the expression to an actual artifact but opened up more associations.

- Bluetooth is rather the odd case in this series. It is neither descriptive nor will it initiate so many associations (except perhaps for Danish historians). It is first obviously a "proper name" and not a common name. For that reason, it seems closer to a brand (much closer than Wi-Fi) which might have been problematic for a standard. However, it was a smart move since it prevented one company from winning over another in the battle to impose a name. Somehow the use of an incongruous name allowed people to create a neutral ground. However, connotations were also present: Blue is a word and a color that was associated with

[38] I use the vocabulary of psychoanalysis on purpose here. I want to account for both discrete behaviors but also scan mode and defining objects specific to designers, or more accurately, systematized by the designers that have to do with intuition and dream thought.

> many technological products at the time (Blu-Ray for instance). Blue supposedly conveyed the coldness and efficiency of technologies. Tooth might have also relayed the idea of something sharp and precise. While it may not have been the reason why it was chosen, it might explain why it stayed. Bluetooth is definitely striking and will elicit questions. But in terms of poetical reception, the "proper name" needs to be "neutralized" so as to be evocative. In other words, we are almost better off without knowing the whole Danish King story, so that we can play on what the words evoke on their own.

Names are also compared to other names and finding the right name might have to do with the need to break from or, on the contrary, to establish a connection with other names and a tradition. Trackpoint (at the beginning called Trackpoint II) was chosen to establish such a connection. According to Ted Selker, it was thought of in relation to other existing products or devices. The naming process therefore focused on establishing the new device within a family, and gave it a sense of continuity, an inscription in the history of tracking devices. Part of the imaginative process was about the lineage of technical artifacts. On the contrary, to choose Bluetooth was to break away from any tradition.

From a conceptive viewpoint, certain names will be more expansive, support more connotations than others: the more technical and limited names are based on the legitimacy of the technical reference; a proper name restricts the connotations to a specific actor; a metaphor opens up more different ways to think about a device or a field of research. This shows that naming is central to a design process. Names expand the scope of the invention and ultimately can support new perspectives and venues of research.

4.3.2 Naming Projects: Acrobatic Acronyms

Giving a name to a project, while seemingly equivalent, is slightly different from giving a name to a technology or a field of research. In my experience, the naming process is a curious one as most participants often do not remember what the name of the project means. One of the reasons is that they are often made of acrobatic acronyms. In fact, people often keep the acronym but change its explanation. "Popeye" was a case in point. The research focus of this project was "peer to peer collaborative working environments over mobile ad hoc networks".[39] How did we ever come up with such a name? What does the cartoon character have to do with ad hoc networks? Questioned about the choice of this acronym, Isabelle Demeure, who was one of the project leaders, remembered: "POPEYE was an improbable acronym: *P*r*O*fessional *P*eer *E*nvironment be*Y*ond *E*dge computing". How many meetings and brainstorming sessions were necessary to get there?

[39] http://www.ist-popeye.eu/

Research projects are currently baptized. Here too, one can witness a variety of inspirations. Some names are more telling than others: PLUG (PLay Ubiquitous Games and play more) as we worked on mobile games sounded both relevant and evocative. "Safari[40]", "Transhumance",[41] were suggestive of mobility, but did not give any real clue to how they were dealing with it.

I remember the process for "ADAM". We wanted to study the uses of peer to peer but also mobile and distributed networks. We wanted to understand how these technologies were spreading in society. In fact, a good part of the discussion was about the object of research itself. Peer to peer was a good starting point but it focused on applications while my team wanted to focus on infrastructures. Finally, the common denominator between the two teams was the fact that we were interested in distributed architectures. That sealed at least 2 letters of the acronym we were looking for: A and D. Starting with that we played on a number of combinations based on the sonority and the connotations of the words (the "voice" of the name[42]): AD + mobiles information: ADMIN that gave the project an interesting connotation of computer management. AD+ evolution of relations: ADER that sounded like the inventor Clément Ader who not only was a major inventor in the aviation field but also for the telephone. AD + ingénierie des relations (engineering of relations): ADIR. But none of these names were "fun" and we finally came up with ADAM that would stand for "architectures distribuées pour applications multiples" (distributed architecture for multiple applications). It was not more evocative than the other options but it combined applications with infrastructure and a biblical name, knowing that at the time there was only one man on the team. What struck me at the time was that the aesthetics of the name mattered: its sounds, its length, as well as the evocations it triggered.

Beyond the private joke, research projects have names for a number of reasons. On the one hand, they serve as a management tool. On the other hand, they belong to the design process in so far as they condense discussions about the research at hand and the identity of the group.

First, when researchers write their proposal, they need to identify the project to have it recorded: the spreadsheet in which the project is to be written includes blank fields to be filled. The space for the name is limited, hence the production of short names that can fit in as well as being nicknames, easy to say and easy to remember. These names therefore primarily help to identify the project, to remember it, to include it in a conversation, etc. Names given to projects can be chosen to be descriptive, to remind people of the roots of or aspirations for the project, or simply

---

[40] SAFARI - Services Ad-hoc/Filaires: Architecture de Réseau Intégré: "Le projet SAFARI propose l'étude, la réalisation et l'expérimentation d'une architecture de réseau intégrée pour la conception, le déploiement et l'exploitation optimale de services dynamiques sur un réseau IPv6 hybride Ad hoc / filaire." "The goal of the SAFARI project is to study, realize, and experiment an integrated network architecture for the design, deployment, and optimal exploitation of dynamic services on a hybrid – ad hoc / wired - IPv6 network".

[41] TranshumanceServices Ad hoc: Réalisation d'une Plate-forme pour les Applications Pair à Pair sur Réseaux Mobiles Ad Hoc.

[42] http://interbrand.com/views/give-your-name-a-voice/

be a memorable word or sound that will stick with people's minds to help the project. They are also used as a group signature: as they qualify the project rather than the technology proper, they represent the different participants, their activity, the results, the time of the project. In my experience, project names try to encompass both the technical features and some pun that commit them to memory. They are descriptive, communicative, and mnemonic words.

Second, from a conceptive viewpoint, they are a poetic production that involves a good deal of imagination,[43] and they also open a specific plane of interpretation. Proper names as pointed out by Searle are "pegs on which to hang descriptions"[44]:

> The uniqueness and immense pragmatic convenience of proper names in our language lie precisely in the fact that they enable us to refer publicly to objects without being forced to raise issues and come to agreement on what descriptive characteristics exactly constitute the identity of the object. They function not as descriptions, but as pegs on which to hang descriptions. Thus, the looseness of the criteria for proper names is a necessary condition for isolating the referring function from the describing function of language.

As defined by Searle, these names therefore benefit from a certain "looseness". They do not describe and give attributes to an object (or person) but still refer to it in ways that open up a whole family of possible descriptions. A pragmatic and literary analysis must then question how they start this family of possible descriptions. In my experience, a proper name both is related to previous stories and starts new stories. When the names are created, actors will make sure that they are relevant to their vision of the project, and when they are told everybody can "remember" different aspects of the project.

To start with, a proper name is related to former narratives. It brings back memories of similar names for each participant. This highly subjective and context dependent relation to names makes it hard for the group to tune their visions. Because, their identity as well as that of the project are at stake, participants who baptize a project have to agree on the narrative connotations that are suggested by the name. Cultural narratives as well as technical knowledge are therefore exchanged during the process evoking different meanings and agreeing on a polysemy that will suit everybody in the group. In ADAM, the combination of Distributed Architecture on the one hand, and Multiple Applications on the other, while originally chosen for evocation purposes, also forced the members of the team to think about the relationship between infrastructures, applications and services. It expanded the conceptualization of these systems within each discipline that was represented; law, economics, information, engineering sciences, sociology, not only because participants shared knowledge but also because they expressed values, told stories, worked on "fitting" expressions.

Second, proper names start a narrative process that recalls the events surrounding the projects. According to Kripke,[45] a proper name stands for the dubbing ceremony

[43] Butor (1969) Bosredon (1997).

[44] Searle (1958).

[45] Kripke (1991).

that accompanies its creation. An important point is that they are context-sensitive.[46] This means that they can take on meanings very differently according to the context of utterance. For each partner of the ADAM project, the identity of the project remained open to interpretation. In particular, we had different ways to talk about it so that our explanation fit each researcher's priorities and the situation where she/he talks about the project. It was not only a strategic advantage that suited everybody's agenda, it allowed new expansions of the concepts. While one of the PhD students involved in the project focused on economic stakes,[47] the other concentrated on the political use of these networks.[48]

4.3.3 *Scientific Logos: A Question of Identity*

The logo as a sign of branding is an interesting case since it brings marketing logic into the world of science. A logo is a "graphic design that a company uses, with or without its name, to identify itself or its products".[49] It supports the consumers' affective reaction to the brand. It is flexible: while a name can hardly change, a logo can evolve, be modernized, to fit graphic trends. In any event, the logo is a strategic asset of a company and is crafted and chosen with a lot of care. It builds brand recognition, brand loyalty, brand differentiation, that are at the heart of marketing consumer products but that make absolutely no sense for a research project!

Therefore, why would any research project spend time designing its logo? If we overlook the marketing strategy that supports brand logic and try to think of how logos are a way to expand beyond the product and even the company, we might learn why research groups still produce them. According to Keller,[50] a logo directly contributes to the knowledge of brands and consequently to brand equity, that is the different values that the consumers attribute to a brand. This is probably a first clue to understanding the role of logos in research. The literature also shows that logos organize the attention because they increase the recognition and differentiation of brands and even trigger an emotional response.[51] This is important in my opinion as it switches the type of legitimacy that we expect from a scientific project and augments it with "emotions".

However, looking closely at the logos for research projects made me seriously doubt that they were as strategic as for a commercial brand. First, the production of logos takes place in a research setting where no one sells a "brand". Second, logos show different stages of elaboration and sophistication. Some will repeat the acronym with some work on the font that can be reminiscent of the referential meaning

[46] Burge (1973).
[47] Musiani (2012).
[48] Huguet (2016).
[49] Henderson and Cote (1998).
[50] Keller (2012).
[51] Kapferer (1997).

Fig. 4.1 Logo of the Safari project

Fig. 4.2 Logo of the ADAM project

Fig. 4.3 Logo of the Popeye project

Fig. 4.4 Logo of the Transhumance project

of the acronym (like in Safari). Some logos took no liberty at all except something that was deemed elegant (Adam). Some are more imaginative: like Transhumance with its mountain, sheep, and chalet, or rather obscure ones like Popeye which does not make the acronym any clearer. Finally, a lot of the logos show a good sense of humor and self-derision. "Transhumance" is a case in point: the metaphor is taken quite literally with sheep on mountains. This sense of humor permeating some logos is probably indicative of a certain identity of the research groups that play the marketing game but do not take it "seriously". It also demonstrates a certain freedom of the research field from hardcore market or social constraints. However, from my perspective, it is more important to understand how they are expansive and contribute to the production of research (Figs. 4.1–4.4).

I would like to look closer at one of these logos that I find particularly successful. It was, in effect, realized by a real designer, whereas, most of the time, research groups try to find somebody – student, someone from the team, family, or friend – who agrees to do it for free! (Fig. 4.5).

Fig. 4.5 Logo of the ILHAIRE project

The project is called ILHAIRE: Introducing Laughter in Human Avatar Interaction, Research and Experiment.[52] The logo was realized by one of the research partners, the French animation company: La Cantoche. They are not a communication company and they do not design logos for a living. Still the company has graphic designers and artists that could work on the project. The whole process is interesting because it is quite different from an ordinary process of branding. According to the project manager, one of the main differences is that the team includes "a large variety of disciplines and interests in the project and there is no formal hierarchy between the members".[53] So, making a decision was more complicated in that respect.[54] Another difference is that the logo is defined in relation to the theme of the project (and not as could be the case in other circumstances to position the "company" within a market). What makes it really precious then is that, according to one of the authors, "it focuses on the identity of the project, the "heart" of the project".

To judge from the exchanges between the different participants, the "identity" was a subject of controversies: social scientists contended that laughter is social, for others it was important to see that laughter was necessary to health and well-being. Engineers focused on the models of gestures and acoustic dimensions. The researchers were also worried that talking about laughter would not be taken seriously. They wanted to make it clear that it was research, which was finally why the baseline was chosen: "the science of laughter".

In all the research projects, the production of the logo is similar to the brainstorming sessions for a name—one of the rare occasions where the different stake-

---

[52] "ILHAIRE objectives are to help the scientific and industrial community to bridge the gap between knowledge on human laughter and its use by avatars, thus enabling sociable conversational agents to be designed, using natural-looking and natural-sounding laughter".

[53] Interview with Justine P., November, 2012.

[54] Yet I find that this absence of hierarchy between disciplines is also a condition of the expansion of concepts as I show in Chap. 7.

Fig. 4.6 Evolution of the ILHAIRE project Logo

holders explain and compare their viewpoints. It is not only the need for a shared culture that structures the discussion. The goal is to design the project, to invent new artifacts made of visions, values, technical and human knowledge and non-knowledge. During the debates, the participants also learn about the process itself, the impact of images, the role of signs and words. The role of the project manager is particularly important in bringing a reflection about the making of a logo. In ILHAIRE, the researchers discussed the fact that a logo is "not an illustration". In particular, it was tempting to use an avatar as part of the logo but then it would work as an example, rather than a sign with a more general purpose. They also worked on how to share values through the choice of a logo: "warm colors", "playful", showing: "The collaboration between different fields of research (engineering + human sciences)".[55]

It is interesting to see how the whole process went from words with denotative values to a process of images with an emphasis on visual culture: from trying to deploy the acronym, to gradually incorporating the culture of cartoon, with references to the field of computer science (Fig. 4.6).

The final logo shows a real mastery of graphics: see for example, the bubble and emoticon that are left open by the frame of the rectangles. But more importantly, it played a role in the representation of the project for the participants. It crystallized a partnership, helped build an identity for the project – a mix of human and non-human actors – with its values and culture, and opened the possibility of a hybrid field of research.

4.3.4 Literary and Visual Productions Supporting Engineering

Despite the differences between giving names to technologies and projects, or designing a logo, I think that the examples in the previous sections shed some light on the convergence between these practices and the process of invention.

Designing logos and crafting names deal with group identities and technological projections. These practices strive to represent the technology or the project without using the usual tools of the trade, like scientific schemas or "proper" technical descriptions. In fact, several cultural trends are embedded within the iconic or name choices. Some of the logos are clearly related to the culture of cartoons. To go back

[55] Working paper, ILHAIRE, FP7, 2011–2012.

to ILHAIRE, it both alludes to emoticons, cartoons, and finally to scientific images through the representation of a sound curve (as is the case with SAFARI). Because they can embed several types of references, they function as a metonymy: they show that the project is at the crossroad of scientific, cultural and literary worlds. Similarly, the expression "Artificial Intelligence" opens up a larger realm of possible research and applications.

However, the question is how to evaluate these productions. Are not some of these productions more successful than others? As we cannot base our evaluation on the brand logic that is largely inadequate in this particular context, success is not related to the relationship of a product with a customer. I will contend that their value depends on whether and how they contribute to the expansion of the technology, field, or project, for various audiences and participants. As we have seen, the goal is not to limit the object to its technical properties. On the contrary, images and words give some interpretive leeway by bringing together several types of references. In a way, there is the same difference between a traffic sign (that would be the strict scientific definition) and a work of art (that deliberately seeks a variety of interpretations). On the one hand, the traffic sign triggers an immediate and monosemic understanding. Through education and practice, we have learned not to think about it but to know about it and act accordingly. It would no longer be a traffic sign if it were freely interpreted and discussed as a text, an image or an original artifact. On the other hand, all discursive and image products are "an open product on account of its susceptibility to countless different interpretations which do not impinge on its unalterable specificity. Every reception of a work of art is both an interpretation and a performance of it, because in every reception the work takes on a fresh perspective for itself".[56] Therefore, these poetical practices explore the cultural meanings of the technology and open it to diverse interpretations.

Primarily, it is interesting to see how these poetic practices are aesthetic and intellectual work that reorganize the knowledge about the technology, field of research, or project, and produce new meaning for the participants. These productions, rather than working on precise attributes, operate what the logician and philosopher of language Saul Kripke, calls the "opening of possible worlds".[57] Even in the case of technology naming that tends to limit the possible connotations, we have seen that the exploration of names becomes an expansive part of a project with results hovering between denotation of what the project is about and connotations of how people want to feel about it. Some names (in particular proper names) also bring out various context-sensitive interpretations. The Artificial Intelligence example shows how the diversity of interpretations might be designed into a name to help the project to continually expand its goals.

Now we need to realize that the "opening of possible worlds" is also related to an operation of condensation and then expansion. Baptizing a project, naming a technology or representing them through logos are all a way to condense an extremely complex and long process and its results in one expression. They are a very difficult

[56] Eco (1989).

[57] Kripke, *Naming and Necessity*.

poetical exercise because they have to contain many attributes, meanings, properties into one single name or image. Participants in this poetic process therefore translate their scientific contributions and understanding for each other. The discussions around ADAM were all about the single words that would encompass different technical and social realities. When the project was defined, it was a constant concern of the group of social scientists to get technical facts right. At the same time, they were determined to show how a technological device or infrastructure lives in and is defined by society and not only by engineers. The word "application" for them conveyed this double meaning of something that is both technical (an app') and social (applied). The translation process is therefore one of drastic reduction but also plays on the potential of words to evoke different contexts and meanings. The process of reduction into one symbolic statement either visual or linguistic is therefore frustrating but relies on the power of certain words and images to open new interpretations through their diverse connotations. It also frees the group of researchers from technical determinacy and opens a dialog with other people who do not necessarily have access to the technical strata but who can deploy their own correlated interpretations. In fact, the condensation process is paradoxically extremely expansive in terms of reception because the best ones, like Wi-Fi or Artificial Intelligence, obviously do not narrow down the field or device but offer many possible conceptual and aesthetic layers. The scientific project is expanded by its aesthetic components.

4.4 Use Case and Story Boards: The Researcher as a Story Teller

Notwithstanding the conceptive usefulness and flexibility of names and logos, the latter remain quite abstract and conceptual. I think that research teams redeploy their imagination in narratives so as to counterbalance this process of reduction and condensation. Beyond baptizing a project, designers and engineers want to anticipate how their findings are going to fit into people's activities. In effect, one of the first deliverables of a research project consists in staging the technology so that the different actors of the project get a sense of what it is all about. Scenarios that describe what the technology will do (as such they have a contractual value) offer a roadmap for the technical specifications and for the production of a prototype that demonstrates the results at the end of the project. The words "use case" or "scenario of use" are often used by engineers and designers alike, but they actually refer to different aesthetic productions that we propose to distinguish and analyze.[58] The study of these texts is part of the reflection on the role of narrative in the conceptualization and design engineering (Lloyd,[59] Turner[60]). For instance, Lloyd shows how

[58] Gentes (2008).

[59] Lloyd (2000).

[60] Turner and Turner (2003b).

storytelling appears to be a central mechanism in the development of a common language in design teams and how it is indicative of good design work. But his focus, as well as Bucciarelli,[61] is more about the understanding of design as a social collaborative activity. Language is seen as a tool to foster design participation. My goal is to understand the expansive properties of narrative productions.

I was lucky enough to follow two teams: one of engineers and one of designers, working in parallel on the same project. This observation uncovered two writing strategies: the engineering research team wrote a use case, where the technology is presented as the main hero in a crisis situation, while the designers staged the technology in context. Both strategies have advantages and shortcomings that are worth noticing. Mainly, the first one concentrates on story-telling and the impact of the "technological hero" on a specific activity, while the other, with a more holistic viewpoint, considers the visual elements of a situation and how the technical artifact blends in an environment. In the former case, the invention solves problems, in the latter case, the technology is part of an environment. Use cases are part of a process of self-legitimization and a communication tool for the research team, as has already been suggested by Clausen.[62] However, I want to show how visual and textual narratives also contribute to a vision of the technology. They belong to the creative process.

4.4.1 Engineering "Use Cases" as Fairy Tales?

Description of the Form of Use Cases

As studied by Cockburn,[63] scenarios – called "use cases" – produced for engineering research projects obey specific narrative and editorial rules. The story telling pattern is recurrently one where a hero – that I call "techno-science" in that it is both a new system and scientific findings – tries to solve people's problems. The goal is a better life for everybody (here the pretext of needs that we mentioned earlier in Chap. 3 makes sense as it gives an almost sacred goal to the research). The hero armed with the technological solutions can eventually overcome the challenges of life.

Seen from the perspective of literary studies, the similarity of the rules for use cases with the structuralist model of narratives is particularly striking.[64] In the 1920s, the formalist Vladimir Propp analyzed traditional Russian folk fairy tales. In particular, he pointed out how "the constant element of the fairy-tale is a function,

---

[61] Bucciarelli, « Between thought and object in engineering design ».

[62] Clausen (1993).

[63] Cockburn (2000). see also http://alistair.cockburn.us/usecases/uctempla.doc

[64] http://vuw.academia.edu/SkyMarsen/Papers/587593/Use_case_analysis_with_narrative_semiotics

independently of who realizes it".[65] Propp's typology was based on the linear and chronological division of the tale into a series of sequences, each sequence triggered by a function. The function is defined by Propp as "an act of character, defined from the point of view of its significance for the course of the action". The plot is carried away by a series of transformations (31 functions according to Propp's typology such as "interdiction", "mediation", "struggle", etc.) from an initial situation to the resolution of the intrigue.

This idea that the tale is based on a series of connected functions was taken up again by the semiotician Algirdas Julien Greimas. Greimas also wanted to demonstrate that there is a "deep structure" within any narrative but he switched from Propp's diachronic perspective to a synchronic perspective. He searched for patterns underlying the narrative and proposed an actantial model presented below[66] (Fig. 4.7).

The descriptive relevance of the model has been criticized because it cannot account for the diversity and complexities of fiction in particular because it occludes the temporal dimension of the narrative and replaces time sequences by logical links.[67] However, it offers a simplified and efficient model of what a story is made of: "a system of oppositions between actants that are bent on achieving a quest".[68]

As we have seen in Chap. 3, research projects are partly based on fictional narrative that present a recurring pattern of people facing ordeals for which the technique invariably provides relief. For example, the scenario "busy man", for the research project SAFARI, features a technology that can help people who take public transportation, to better manage their time and react more efficiently to disruptions of traffic.[69]

> I arrive at the station. My train is leaving soon. Upon my arrival near the station, my handheld knows nomadic services offered by the station. I click Trains, Departure Board I check the time of departure of my train and find the platform from which it leaves, followed by a map to access the platform.[70]

Even though the story is based on members of the team's personal experiences, the text removes the marks of a subjective point of view and proposes a generalized "objective" transcription of the experience (the text of our example is written in the first person but it is a universal "I", not a subjective personal experience). Thus, the characters are not individuals but 'agents'. The focus is not their psychology or any background in their jobs or skills. The "actors" are described by their movements and gestures as if they were part of a mechanism. Conversely, techniques become heroes in their own right. The manuals that help write use cases explicitly refer to

---

65 Propp (1968).

66 Greimas (1983).

67 Sack (2013).

68 Hebert (2006).

69 ANR SAFARI, 2003–2005, Working Papers ENST-SNCF, 2003.

70 ANR SAFARI, 2003–2005, Working Papers ENST-SNCF, 2003.

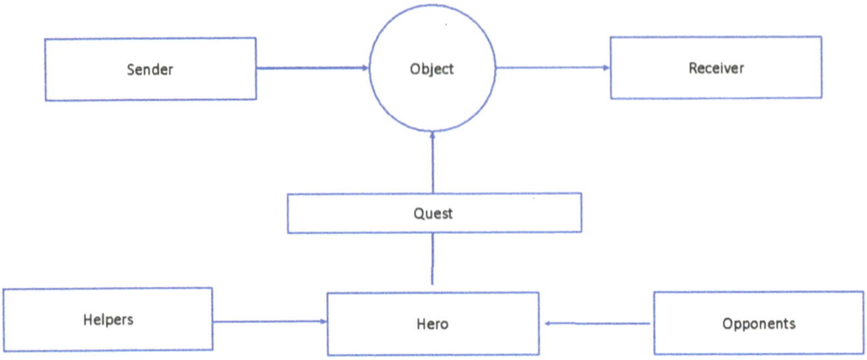

Fig. 4.7 Greimas: actantial model from structural semantics

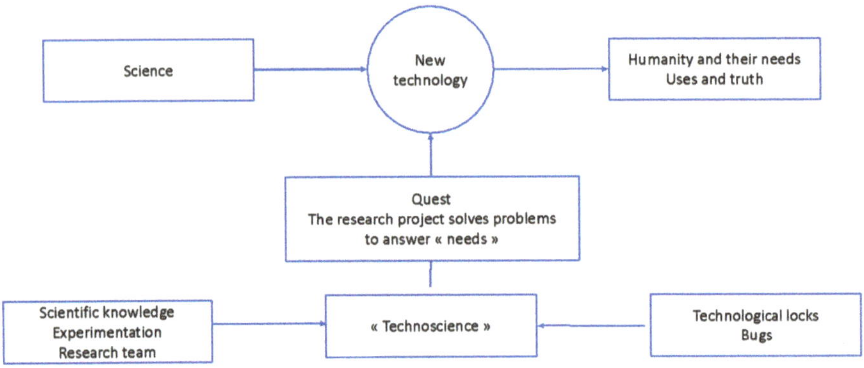

Fig. 4.8 Actancial model of engineering research projects

this behavioral, almost, mechanical dimension of the actors. "Actor: something with behavior, such as a person (identified by role), computer system, or organization[71]". In my experience, the figure below represents the general techno/scientific model that most engineering projects use (Fig. 4.8).

The use case is often also written like a flow chart, so that it can be divided into abstract units that are eventually distributed in a formal grid that lists the "actors", and gives a title and a number to the different sequences. Each sequence has a purpose: it must "translate" an action into a technical feature: use case diagrams. Each feature of the narrative must be described in terms of processes: entry into the device, creation of contents, access to tools, management constraints, etc. This table will eventually lead to the specification documents. For instance, the Computer Science Department of the University of Helsinki has a script for a role play in mobility. We note that the "actants" are limited to an extremely schematic figure

[71] European Project Popeye Working Paper: Introduction to Scenario Collection Methodology and UC, 2006.

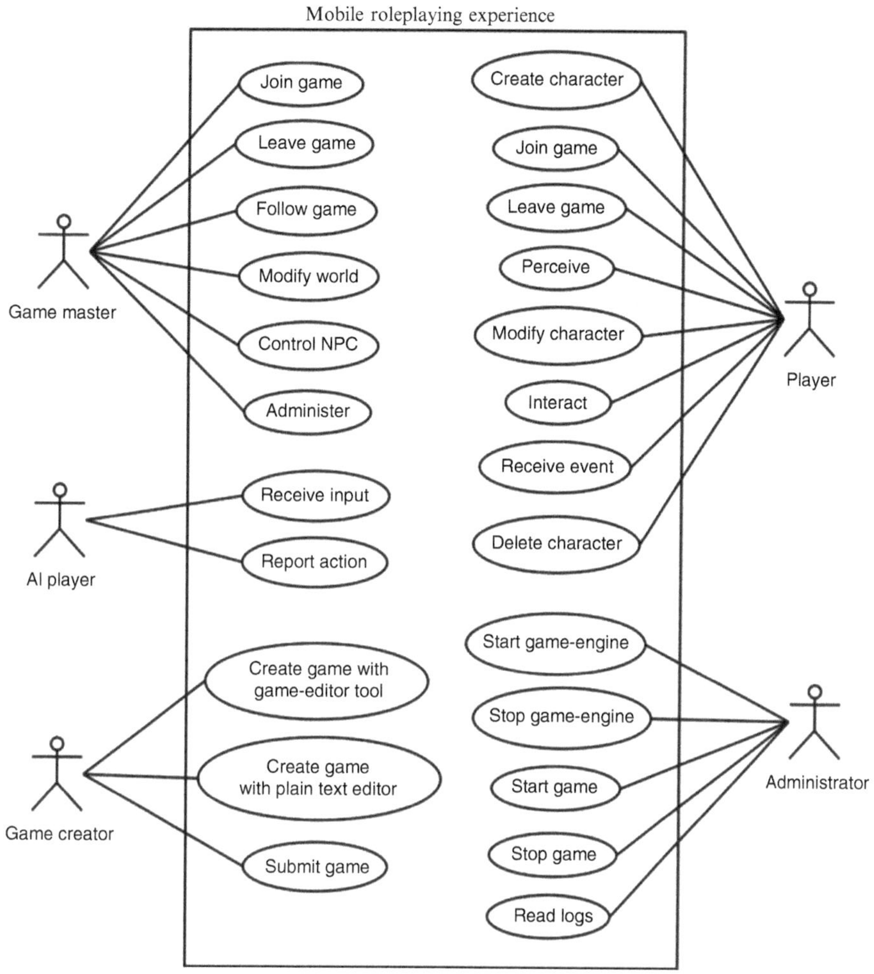

Fig. 4.9 Example of use case diagram for a mobile game Helsinki (http://www.cs.helsinki.fi/group/mobi/mrXdocs/usecases.html)

with a title: either master, game player, administrator, … These terms reinforce an actantial vision of the stakeholders in the scenario: they are " elements of the system " defined by their task (Fig. 4.9).

These diagrams are extensively used in the community as a simple search on Google Images shows. They can be found in contexts of education, research, association, companies. Once the actors are identified they are organized on each side of a list of actions. The actions follow a chronological order but also helps formalize the structure of the technology as seen in the use case diagram from the project Transhumance (Fig. 4.10).

This format that focuses on "actants" who activate features is not without consequences on the representation of the emerging technology. First, the technique is

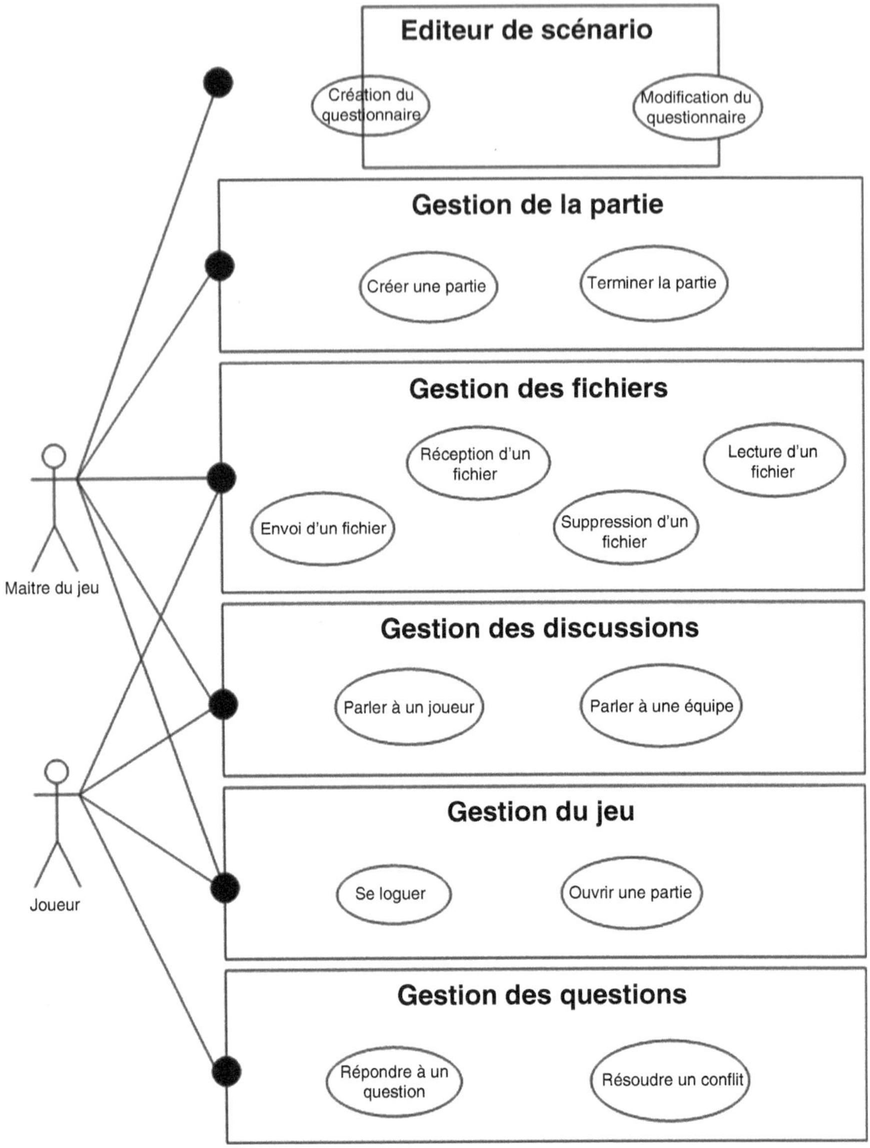

Fig. 4.10 L4.01. Use case diagram of the hunt game. ANR- Transhumance – RNRT – 2005

presented as a tool, an extension of the hand and the will that animates it, performing a physical or mental activity uncorrelated to culture or society. The object is transparent to the action. It is also advised not to describe the how but the what, that is to say the result of the action. Cockburn recommends: "They should ignore when possible "how" an interaction between the actors and the system is performed and concentrating on "what" they do, i.e., which valuable result they produce".[72]

This "transparent" object serves to increase the power of the agent, his/her influence on the world despite the obstacles. For example, in the European project Popeye, the researchers envisioned a computer game that supposedly took place in a castle in ruins, which naturally contributes to the atmosphere! But more importantly, the protagonists would have no access to the Internet and no electricity. The technical platform on an ad hoc network solved the problem because each participant and spectator could share the same environment and see the evolution of the game on his/her computer or PDA.[73] This extended sphere of action of the agent was justified by a single drama resource: facing a critical situation, the technology and its carrier found a solution and the situation could return to normal (ie. a perfect balance of activities). The agent's quest is the good of "users". There are opponents (technical constraints) and adjuvants (for instance, the *Optimized Link State Routing* Protocol (*OLSR*) that is an IP routing protocol optimized for mobile ad-hoc networks).

Expansive Writing Tools: "Writing Leap", "Dramatic Spark", "Staging Moment", and "Flow Montage"

Are these use cases caricatures?

Yes! in the sense that the situation – with its limited definition of actions, absence of subjectivity, absence of context, and technical features at the center of the narrative – boils down to a set of technically actionable operations.

No! However, I contend that use cases also have expansive properties since they help participants to give birth to a project. Looking at the writing process and its impact on the project, I mainly find four different properties that I call: "the writing leap", "the dramatic spark", "the staging moment", and the "flow montage".

First, use cases help researchers detach themselves from their personal history to assume the role of narrators. To start with, a project is grounded in researchers' subjectivity. For instance, in Transhumance one of the project managers had a strong impact on the outcome of the project because she defended certain values that were finally embedded into the demonstrator: she wanted people to cooperate and refused to organize a strong competition between players. She also wanted cultural contents. But the written documents operate a passage from subjective experience rooted in the personality of the researcher and her interactions with the team to objective fiction. Scenario writing is about rewriting elements of personal history into a fiction of science. There is a need to switch from the realm of experience to the realm of text. **It is the "writing leap", where the individual person becomes a writer and narrator.** It means that researchers should be more confident that they can share their experience but also acknowledge that as such their experience is transformed into a text where they have more freedom to collect and aggregate other material to build the fictions.

[72] Cockburn (2000).

[73] PDA: Personal Digital Assistant.

Second, use cases describe the need for the future technology which means that they find a way to create a negative photograph of what the technology is about. The narratives must create an "aporia" that the technology alone is able to solve. Writing a scenario – like the computer game taking place in an old castle that I mentioned above – is about creating the conditions that are going to start the whole process. Scenarios must therefore capture the "essence" of an emerging technology as it topples the existing state of things. For example, the game scenario for adhoc networks that considers the tradeoff between energy management and the number of exchanges between players, is revealing of the challenges of the technology proper. Indeed, the management of this compromise – "hidden" in devices powered by electricity – is apparent on a mobile device whose batteries are limited. The scenario is therefore deemed relevant to the research. *Conversely,* scenarios that replicate known functionalities that could be supported by other technical platforms are turned down. For example, one of SAFARI's scenarios was critized because nothing in the service compelled the architecture to be distributed nor mobile. For the team, the scenario had to show more clearly what was meant by a mobile and decentralized architecture. **Finding the right set of circumstances that sets off the technology is the "dramatic spark".** For the research team, it means that the important aspect of the narrative is not necessarily its implementation as would be expected but how the story reflects on the technology.

Third, scenarios help researchers to evaluate existing techniques. They give a clear view of what to develop or rearrange so that the activity that is described can take place. The scenarios list all the hardware and software that will be used as well as how they are going to be challenged by the imagined situation. All the materials and tools are laid out. **It is the "staging moment".** As such, the staging moment participates in the architecture of the system, but it could include other props that make it more significant and therefore could tend towards an ecology of the invention.

Finally, stories and "use case" diagrams represent an organization over time. They represent a task as a series of individual and collective actions. The temporal sequence of actions is visualized. **It is the "flow montage".** The flow montage in use cases has two distinct features compared to montages in films. First, it supposedly admits no ellipse in time: it gives the illusion of a perfectly controlled and exhaustive series of actions (that the actual use also often contradicts). Second, it is totally retroactive. One can move forward or back without inconvenience. A certain vision of time is therefore laid out by the use case: linear and easily divided into micro sequences.

These four properties that structure the writing process are all expansive properties of writing in that they turn the individual experience of intuition, ideas, and possibilities into a social and aesthetic process that shapes and increases the set of possible options. As we have seen, the writing leap helps the individual researcher put her ideas into words and then reflect on them. It is a moment of appropriation and critical distance because through the scenarios the values appear more clearly and, therefore, the technical project can be discussed not only on technical terms but also on cultural and ethical terms. The dramatic spark also has maieutic properties as it offers a view of the essence of the technology. The narrative of a situation sets off the goals of the technology and its ultimate originality. The dramatic spark expresses the unique characteristic of the invention. For the audience, the originality must be obvious and therefore the invention is implicitly compared to other similar technical situations. It is a difficult poetic trick since the actual scientific work is of course based on continuities and not only breakthrough findings. But the authors have to step back from their day to day scientific experience and find a way to express the intrinsic difference. The staging moment, as we will see in more details in Chap. 5 on design as composition, is the gathering of all narrative elements that are going to flesh out the dramatic spark. The definition of the invention depends just as much on all these details, scientific facts and artifacts, atmospheres, actors as on the legitimization process. Each element actually redefines the whole technology giving it all kinds of different facets. Finally, the fourth property, the flow montage technique, brings all these elements of composition into an actual timeline. The chronology forces the authors to reconsider the different activities of the actors, the deployment of the technology, the fine sequence of interactions with the interfaces as they are used. Considering what is at stake in the whole process, and what it brings to the invention, is therefore a necessary step to understanding these practices not only from the perspective of communication and management but also from the perspective of creation.

4.4.2 The designer's Storyboards as Theater Sets?

Even though designers also speak of scenarios, they produce very different documents. Among the various productions (texts, sketches, 3D visualizations, diagrams) some structure a story through what I prefer to call "storyboards". These storyboards are videos of places in activity and actors manipulating the device or photo montages introducing objects in space.

I use this term "storyboarding" because designers, as film directors, scout locations and one of their main question is: "how will this technology fit in this particular place?" When trying to figure out mobile ad hoc applications in railway stations[74] and in the Underground,[75] designers first observed how stations were organized, how people moved, used the station, read the various media of information. Photos of circulations, urban furniture, signage, people in action were taken (Fig. 4.11).

In the SAFARI project that was dealing with ad hoc networks in railway stations, two storyboards offered to build interactive cubes in the stations and used photo montage to show how these new objects could be inserted in the station and what kind of interface they would show. In "Lost and Found," they represented a form of community and its activities, in order to work on the sociability of people, and create opportunities for discussions. In the project "Make a wish", they showed the

[74] Safari, 2002–2005.

[75] Smart Cities project with ENSCI and Bell Labs.

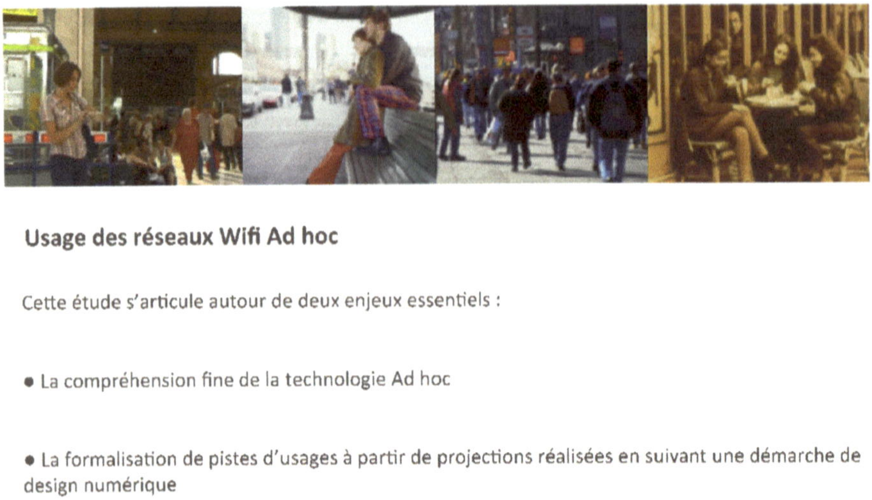

Fig. 4.11 Jean-Louis Fréchin – ENSCI. Projet Safari-RNRT. 2003

community of contributors. In their pre-figurative sketches, they represented "screens sculptures" in the space of the station as if they were already there, as if they were part of the scenery. They therefore relied on the architecture of the place and worked on the consistency of such interactive sculptures within these locations (Fig. 4.12).

Building a sculpture was a way to contest the idea that these technologies are immaterial. The projects were an indirect criticism of the concept of ambient technologies. In fact, designers wanted to show and test how these technologies structure different places that people inhabit or attend.

In the work of designers, the location is therefore not an excuse for action as in the engineers' use cases, but a complex element that reveals the aesthetic qualities of the technique by juxtaposition and contrast. Like a theater set, these pictures give meaning to the play that is about to be performed.

First, the actual location, architecture, other objects, circulation, and people, help define the technology. Technology does not solve a problem but creates a new situation. These visualizations offer a new "distribution of the sensible" (Rancière[76]) that is a reorganization of the different components and attributes of a location. With this type of situated representation, designers can consider the meaningful relation between the object and the space in which it is located. The object is a "prop" which completes the scene and the characters. It fits into and is made credible by an atmosphere and by a social environment. It therefore fits more than the activity of the user. It fits (but could disturb as we see in Chap. 6) a whole situation of use.

As an example, Anab Jain, founder and director of the design firm Superflux, a TED fellow, and an inspired futurologist, who worked with Dr. Patrick Degenaar at

[76] Rancière (2006).

Fig. 4.12 Jean-Louis Fréchin et alii – ENSCI. Projet Safari-RNRT. 2003

the University of Newcastle on artificial retinas, created a film: "Song of the machine[77]", where the camera operates in two main ways (Fig. 4.13).

In the subjective camera mode, we see what the visually impaired hero sees through his visual device. Our own vision of the movie is affected by the choice of visual functions by the character. In the omniscient narrator mode, the video shows us how the character runs his daily life with his visual prosthesis: he has an appointment, he has lunch, goes for a walk. The spectator experiences the technology inside out: the "glasses" are an object that we see on a table, or on the nose of the protagonist, like a prop. And when we do not see them, it is because the spectator wears them too in first person visualization mode. In both modes, the movie offers a holistic vision where every element, and in particular device and location are inseparable. They "work together".

The object also makes sense in a specific situation. Duchamp's powerful invention of the ready-made points precisely to the fact that the urinal called "Fountain" and placed in a gallery, transforms the meaning of the urinal but also the role of the museum and that of the artist. As the art historian Nicolas Bourriaud notes, Duchamp uses the museum as a film on which the object is "imprinted" and therefore, in the same way, changes its nature from useful artifact to art work.[78] Similarly, the tech-

[77] http://superflux.in/work/song-machine

[78] Bourriaud (1999).

Fig. 4.13 Anab Jain, "Song of the machine", 2011

nology once introduced in the field changes the scenery and at the same time is influenced by the location.

Finally, storyboards raise the question of what the technology and its user will be able to incorporate from these places and technologies in terms of personal identity and social interactions. The notion of "place" as understood by the anthropologist Augé[79] and analyzed by HCI specialists[80] is relevant in this context of design. The analyses of photographic and video documents show the importance of inhabited and practiced places in the development of technology by designers. When the designer Aude Guyot thinks about developing applications in Senegal where maps are almost non existent (in 2009) or inadequate because of the ever changing topography of the city, she decides not only to spend a long time of observation in Dakar but to write a book about mobile practices where photographs of iconic behaviors or urban details such as advertisement in the street play an important part of contextualization.[81] Her service, once designed, is immediately staged in the same environment with "actors" who stroll the same streets and manipulate their cells in places where it makes sense for them to do so either close to a bank or a touristic place of interest. Designers not only consider the geometry, volume, light, materials, but also the way people and technology interact with these elements. The technology and their users co-evolve with these places both in terms of identity, and relations but also history. History here "is defined by a minimum stability. It is historical in so far as those who live there can recognize markers that do not require to be knowledge objects."[82]

---

[79] Augé (2009).

[80] Harrison and Dourish (1996), Turnera and Turner (2003a) and .Dourish (2006).

[81] https://issuu.com/audemai/docs/instantanes_senegal

[82] Augé, *Non-Places*. p. 71.

The corollary of this meaningful relationship between object and place is the relations between people and how they are represented by the interface. Much attention is paid to the problem of figurative or abstract representation of people in both network and in the actual place. Some mock-ups of the SAFARI project were based on avatars similar to those of virtual worlds as Habbot Hotel,[83] others represented the users by simple abstract points. In the designers' staging, the management of proximity, gaze, possibly unveiled anonymity, immediately appeared as a fundamental issue of ad hoc networks, while the question was discarded in the engineers' scenarios because the focus on actions/functions largely leaves aside the question of representation. In other words, the graphic work on the interface also reveals traits of the technique that escape the purely textual narrative scenarios.

The advantage of story board techniques is that they support different levels of details. One may want to look at the bigger picture or get closer for a discussion on details. Discussions turn from micro elements to macro elements of the project. For designers to maintain these levels of formalization – micro versus macro – is crucial. This makes it possible to "develop the architecture of the object or service without losing" sight "of screens that users manipulate".[84]

4.4.3 Narrative Cultures

In these projects, two narrative cultures with fundamentally different conceptual effects appeared. On the one hand, engineers' productions aimed at increasing the transparency which leads to a formal abstraction of the technical system. The story filters events, and keeps only the ones that seem to reveal the essence of the technical system. As writers, they work through progressive edit checks of the technology, towards a hypostasis as the only way to define the object. On the other hand, designers introduce a whole context, teeming with objects and characters. For them, the technology is a prop amongst other elements of the context: hence the display of daily activities in a home, a train station, or in an urban setting. The storyboards show the object while integrating it into a coherent whole (Fig. 4.14).

Engineering perspective	*Design perspective*
Use case: "actantial" schema:	Story board
technology as "agent"	staging and spatialization of the technology as « prop »
Media: narrative text and diagrams	Media: photo or video montage

Fig. 4.14 Summary of staging techniques

[83] http://www.habbo.com/

[84] Interview of Deborah Elalouf, designer, by the author January 2001.

Beyond their differences, the analyses of narratives by Ricoeur[85] and popularization by Jeanneret further help to identify the creative work that takes place in these documents. As explained by Richoeur, "the 'emplotment' is the operation that draws a configuration out of a simple succession of events."[86] Use cases and storyboards rely on a proper grounding of the story in action. What is described makes sense because we recognize activities. At the same time, they are textual productions. We recognize them as such, fictions carrying values that contribute to the imagination of the technology. But the narrative requirements also project the technique in the users' hands, already incorporating it in the users' lives. A story, by necessity, involves how the actions must be perceived, and how consequently they are part of a particular symbolic system. In this respect, use cases and storyboards complete each other. First, scenarios present an acute awareness of activities in time and, second, storyboards embed these actions primarily in space, both real and symbolic. These documents partake of an effort to create an imaginary universe consistent with the emerging technology. They are basically techniques of socialization of the technology that can be shared by different stakeholders because they complete what Yves Jeanneret calls the "three programs of popularization":

> They tell a story strong enough to withstand a full representation of the unknown (narrative requirement) they articulate the elements of a judgment on a matter of interest (argumentative requirement) they provide a body of knowledge intelligible and compatible with the reader's experience (educational requirement).[87]

The key words here are of course the staging of the unknown: the invention is yet to come or it is unheard of but the scenario has to create a probable simulation. The analysis also shows the importance of demonstrating the values of the invention or at least the way it will question the users' values. Finally, the picture has to be consistent internally (all the elements of the plot have to stick together) and externally (the elements have to elaborate on a certain knowledge base). In the next and concluding section, I want to establish how the different traits described in this chapter constitute a particular genre that I propose to call "expansive fiction".

4.5 A New Genre: "Expansive Literature", Suspension of Disbelief and Future Building

Science fiction is often quoted as a source of inspiration. Specifically, for our project with RFID and artifacts in the museum of Arts and Crafts, one of the leaders of the project, Eric Gressier-Soudan often mentioned the story "The Age of the Pussyfoot" by Frederik Pohl (1969) that influenced his view on the project. The point Eric made was that the device in the book could be made today: looking like a scepter, it was

---

[85] Ricoeur (1990).

[86] Ibid. p. 102.

[87] Jeanneret, Ecrire la science. p. 322.

in fact a portable computer that could be interfaced with objects and places. From this story, he got the vision that would turn the telephone into a sort of magic wand. Another member of the project, Emmanuel Zaza, game designer, also explicitly referred to Captain Kirk in Startrek using the "Communicator". Science fiction can therefore be a "ready to use" scenario that gives a certain type of coherence to a project.[88]

4.5.1 *Fictions of Science as Reflection on the Present Times*

But is science fiction different from the scenario of use that designers and engineers write and how so? Frederic Jameson tries to answer this question when he wonders from a literary point of view if and how we can imagine the future.[89] In fact, literary analysis of science fiction points towards the deep social functions of this genre. In particular, some critics say that science fiction is the place where the hidden workings of a society finally emerge as the subconscious of the community. The fantasy that a society has about itself is expressed through sci-fi scenarios.

Reading such stories one could detect:

> behind such written *traces* of the political unconscious as the narrative texts of high or mass culture, but also behind those other symptoms or traces which are opinion, ideology, and even philosophical systems—the outlines of some deeper and vaster narrative movement in which the groups of a given collectivity at a certain historical conjuncture anxiously interrogate their fate, and explore it with hope or dread.[90]

Similarly, use cases and storyboards embed values in the projects. At least two meta-narratives are at play in our examples: the apology of technical progress and the definition of the "homo communicans", the ideal communicational human being. As we have seen, use cases and storyboards in information technologies are about the dream of an information society that strives to handle our ordeals but also that becomes part of the aesthetics of our everyday life. The focus seems quite narrow but it actually tackles a number of issues. In PLUG, that was developed for the Museum of Arts and crafts in Paris, the design team questioned our relationship to culture and in particular technical culture. How, in an age of multiple media, can the museum still be a place where people go, learn and enjoy what they see? The design project was to invent new media to introduce the artifacts and to invent a way where visitors could interact with each other too. In SAFARI, urban life and transportation were at stake. In particular, it was apparent that the stress caused by long and hectic commuting was at the center of the designers and engineers' concern. Working with the national railway company, SAFARI was also an attempt to create new forms of social intercourses in public spaces – namely railway stations – that would not solve

[88] Reeves (2012).

[89] Jameson (2005).

[90] Jameson (1982).

the problem of commuting but would make it more humane. Transhumance was also about culture but more deeply about users' responsibility in an ad hoc network. Are we ready to share each other's bandwidth, with whom, at what cost for our own quality of service? Whom do we trust? What are the mechanisms of trust that we can invent?

This seems very close to what the writer Gordon Dickson describes in his anthology of science fiction. He remarks that science fiction paints scenes that describe the possible consequences of our current problems and suggest possible solutions.

> « Science fiction is, in fact, essentially an unstructured think-tank in which authors of differing points of view can paint differing solutions or eventualities suggested by present problems or situations. As a litterature it is favorably designed to act as a vehicle for ideas or arguments, to be a seed bag for a philosophical fiction. »[91]

Sci-fi is therefore as much a comment on the contemporary state of affairs as a work of imagination. By a turn of the narrative, our present becomes the past in novels or texts that deal with the consequences of our present. As Jameson points out, the science fiction novel is a way to approach and deal with our present indirectly. It uses special narrative traits to defamiliarize us from our society. While researchers are not writers, they certainly share this ongoing philosophical debate and more acutely so perhaps because they consider their responsibility in bringing about solutions to worrying situations: the future of culture, the stress of urban life, the redefinition of communities.

4.5.2 How to Produce Good Narratives?

However, the question raised by these narrative styles is their quality. Science fiction novels and stories are analyzed and evaluated by many critics who discuss their literary merits. These discussions are necessary for expectations to be discussed, criteria to be clarified, trends to be perceived. In an interview, Alex Pang, futurologist working at Stanford Research Institute, aptly pointed out that one of the difficulties was the lack of comparative studies of these texts that are mostly produced for the industry and thereafter are not published[92] or, as in research contexts, that are not considered as worthy scientific productions and therefore discarded. Fortunately, designers tend to keep these productions for us to judge. In any event, a research goal and organization should enable the researcher in design to collect and compare these texts, placing them within a historical perspective that would help identify poetic practices, different genres and styles from different actors over time.

However, to compensate for this lack of comparison and critical analysis in research fiction and prose, I think that some lessons can be learnt from what literary analysis says about science fiction. In particular, it emphasizes the fact that there is

[91] Dickson (1975).

[92] Interview of Alex Pang by the author, Novembre 2012, see also http://www.future2.org/futures/

a tension between the narrative qualities and the descriptive qualities. Either fictions are driven by the plot, the turns of events, the surprises and resolutions, or they focus more on the re-constitution of atmospheres, people, environments. In the descriptive paradigm, qualities lie in the careful, realistic, description of the world. Every element serves the purpose of a consistent story. If travelling through space is made possible, this technological achievement must be coherently related to other aspects of the worlds that are described. For instance, it should impact the economics of the world and explain the fact that spaceships colonize other planets or do commerce with them. Contrary to Fantastic literature that always questions the status of things described – whether they are from magical sources or from the imagination of the characters – the science fiction world must be "logically" explained. Noah Raford who did his PhD at MIT on "Large scale participatory futures systems"[93] claims that a counterexample to that is "A Day Made of Glass" videos made by the company Corning.[94] These videos, he says, were "wildly successful", but they were also criticized because they lacked "the most basic considerations of causal relationships and interactive effects."[95] The chain of causes and consequences should not be bypassed. A successful sci-fi novel or movie brings both plot and world together. As far as engineers' narratives and designers' productions are concerned this tension between a good plot versus a beautifully crafted future world, seems to me to be part of the equation. Of course, the tension is solved differently since they are distinct productions that fulfill other functions than merely creating a good fiction. In any event, the way research teams evaluate their production points towards two sets of requirements:

- the definition of a situation that makes the use of the technology probable (with agents, locations, activities that are affected by the technology)
- a near future setting that makes it urgent to develop the technology (with references to and a diagnostic of our present time)

The incorporation of a near future in a scenario means that the text has to orchestrate probability. In a way, research has invented a futurist prose, a genre that we could qualify as "expansive literature" that organizes the "cooperation" between research, that is investigation and experimental data, and fiction "the creative activity of the writer's imagination[96]".

What is at stake is the suspension of disbelief: the stories, diagrams, photo montages all strive to open a new design space. As much as they can, the scenarios use the tricks of the trade: realistic setting, believable characters, coherent script, careful montage. The aesthetics are devoted to support a consistent narrative. In this situation, the new artifact is a make believe "prop" that appears as a real industrial object. The quality of the literary production is therefore backing the probability of the whole fiction.

[93] http://fr.slideshare.net/noahraford

[94] http://www.youtube.com/watch?v=6Cf7IL_eZ38

[95] http://news.noahraford.com/?p=1625

[96] Ricoeur (1990)

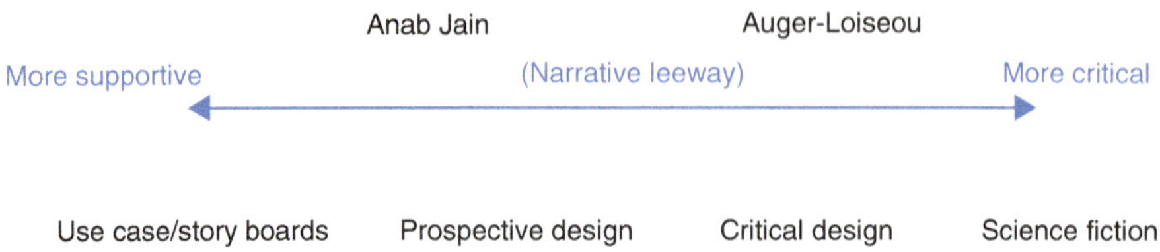

Fig. 4.15 Graduation of "critical ity" in narrative plots

4.5.3 Beginning of a Typology

Another characteristic of these documents is their situation on the line of literary productions: they may question the situation but they are not critical texts of the future to be. Industrial and academic research looks forward to a better future and do not support a catastrophic version of the evolution of technology and society.

In a striking article in the Atlantis, pointed out by Alex Pang, Eric Garland, was precisely explaining the limits of futurologist work because of a lack of critical honesty: "I am not quitting this industry for lack of passion, as I still believe -- more than ever -- in using good information and sophisticated analytical techniques to decode the future and make decisions. The problem is, the market for intelligence is now largely about providing information that makes decision makers feel better, rather than bringing true insights about risk and opportunity".[97]

Research texts are therefore more utopian than really exploring the alternatives, either positive or negative. A comparison between designers' texts can point to four strategies:

- the text supports the best that the technology can bring about. The scenario is closely related to marketing.
- it toys with the technology by placing it in a positive environment but leaving some room for questions.
- it challenges the technology by pointing out the possible consequences, including pessimistic prospects. Some designers from Royal College of Art (RCA) in particular, produce such nuanced narratives.
- it is totally free to produce either utopias or dystopia, because it is within the literary field of science fiction (Fig. 4.15).

To better feel the difference between different types of narratives, we can compare the work of the designers, James Auger and Jimmy Loiseau, on the one hand, and Anab Jain, on the other.

James Auger and Jimmy Loiseau are designers who teach at RCA and produce objects that stir debates. In particular, in the line of H. G. Wells, they came up with a design idea that was picked up by the media and that led to what James Auger considers as a design method today. James and Jimmy had the idea of combining

[97] http://www.theatlantic.com/international/archive/2012/04/peak-intel-how-so-called-strategic-intelligence-actually-makes-us-dumber/255413/

body implant and communication technologies and came up with the idea of a phone tooth:

> Just chatting one day I remember exactly the time and place this idea of the telephone just entering the body suddenly came up. It was a mix of telecommunications that I was really looking for. An implanting technology to speculate realistically how the first implant technology might realistically go into the body.[98]

He qualifies their work as building "alternative presents":

> So the majority of our products are there to exist as speculative future or for alternative presents. Alternative presents are just a different configuration to the way that how things are, that allows us to critique them and to imagine them. A different way of being.

James makes it clear that he is not doing commercial design in this particular instance. He recalled a turning point in his career when, as he worked to redefine what the phone could be for a major phone company, he got negative feedback because the company could not implement a radical change. But more importantly he wants design to be part of a reflective process.

> The philosopher Neil Postman[99] was talking about this change. He says this technological change is ecological. He talks about the world being like a beacon of water, and the technology being like a drop of red ink. So you put a drop of red ink, that diffuses into every different part of society, and you can't take it away again. So you could say, mobile phone technology have done that, they brought those things into the world, you can't suddenly take them away, like that. But I can do that, and it allows us to create a new way of philosophizing through these design objects. So that's my motivation. I do strongly believe it is a design process, predominantly through methodology but its goal is different from a mainstream design process. Its goal is to question. Its goal is to open the mind. Its goal is not to add more objects like these to the world. It's about understanding truly the impact that they are having, because no one is truly questioning that at the moment. You know, we all put Apple on a pedestal for creating these beautiful products, but we don't look at the control that they have over us. It's the most radical shift in human history possible, but we are all bobbing along without any thought to what this means.

James' vision is therefore weary of technological progress and definitely more critical than use cases and storyboards. This makes sense since he is not in the process of supporting these technologies but rather of stepping back and taking a closer look at what is at stake. His work is therefore at one end of the spectrum of the fiction of science. James and Jimmy create challenging objects that are both and at the same time credible and weird. The audio tooth is a case in point (Fig. 4.16).

[98] Interview of James Auger by the author, April 2011.

[99] Postman (1993).

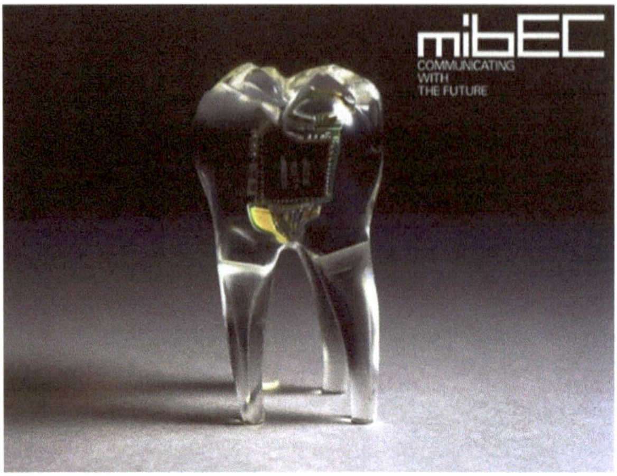

Fig. 4.16 Auger and Loiseau's "Audio tooth implant" (2001) http://www.auger-loizeau.com/projects/toothimplant

Fig. 4.17 Auger and Loiseau's "Audio tooth implant". US Time magazine front cover (2002) http://www.auger-loizeau.com/projects/toothimplant

It was picked up by the press and blogs as a first quality invention. The whole purpose of the production, that was to start a debate about our numerous communicational prostheses and how intrusive they can become, actually worked perfectly well since every one discussed the pros and cons of the invention (Fig. 4.17).

James is careful to point out that he does not think that conning the press is a design strategy as such, and many of his and Jimmy Loiseau's artifacts have also attracted lots of attention from the press without the initial misunderstanding. But the point that I would make here is that the project actually puts into place three strategic elements: first the probability of the technology, then the near future, since the "invention" is identified as a prototype, something that might happen but is not there yet, and, consequently, a debate on how these things might impact our present. Their strategy is supported by the aesthetic and pragmatic qualities of their projects that always have some shocking aspect: intrusive, violent, physically challenging. The body is at the center of their reflection as a way to show that these technologies change us in deep ways.

Anab Jain describes her work as "design futurescaping", which is "using design methods like storytelling, experience prototyping, making scenarios tangible, and talking to people on a daily basis, to inspire and influence prototypes for the near future."[100] A major focus of her company, Superflux, is to design for the '**imminently probable**' – exploring the design possibilities and near-future implications of emerging technologies on people, culture and our environment. "Song of the Machine", that we analyzed earlier in this chapter, is a case in point. It has an ordinary setting in London. The activities are mundane: making tea, reading the newspaper, walking through a park, taking the underground. Developing the project, Anab's team faced several questions:

> How might you choose to 'compose' your vision of the world? How would that affect your sense of the world, and your place in it? What would it mean for your memories? Your dreams? How could you modify your environment to capitalize on these extended senses?[101]

The pitfalls of this kind of project are quite clearly identified: the scenario could be some kind of story where a super hero with extraordinary vision could be doing incredible things. Examples abound in TV series and movies of such augmented bodies (like*The Six Million Dollar Man – 1974-* based on the novel by Martin Caidin, *The Cyborg*[102]).

> It would have been easy to allow the film's narrative to mushroom, obscuring the affordances and possibilities of the technology. Instead, we tried to keep the story simple. A mundane narrative, ambient cityscape, and offbeat score helped set the scene for exploring the electromagnetic spectrums and visual 'channels' of our user's sensory perceptions. Ultraviolet, infrared, and augmented reality extended his world, with the digital artifacts and noise hinting at the technical limitations of low-res vision.

Compared to Auger and Loiseau, her production is less frightening, less challenging for the body, less extraordinary. The relationships between the devices that she is coming up with and the people are "modest" in that they do not stage death, sex, or anything that can be considered as shocking by certain audiences. Her productions do not undermine our set of values. They rely on very slight changes of life style that are made perfectly plausible. In a way, it leaves more leeway for the spectator to imagine other scenarios.

Despite their differences, both designers and engineers have created a new genre because their videos and texts take place in between, on the one hand, science fiction proper (imaginary future) and, on the other hand, documentary/scientific productions (certified present). Their productions can be more or less critical or

[100] http://fellows.ted.com/profiles/anab-jain

[101] http://superflux.in/blog/song-of-the-machine-in-depth

[102] http://en.wikipedia.org/wiki/The_Six_Million_Dollar_Man

challenging of the technologies to come but they are never pure fiction. In other words, designers and engineers have created a new genre, "expansive literature" to deal with the complexity of today's scientific and technological world by producing a whole gamut of projections from the optimist and supportive to the pessimist and challenging.

4.5.4 *Staging "Change"*

Why does the study of narratives matter for invention, engineering and design? I think that this last part of the chapter showed that narration is a flexible material that can expand the way scientists and creators look at things. Writing the future is part of the episteme of engineering science and design, something that Anne-Françoise Schmid, epistemologist, and philosopher of science, investigates when she describes the imbrications of scientific knowledge and fictions in the making of science (see chapter 7). As we have seen, this new genre of expansive literature gives a space that allows for the creation of entirely new paradigms that are nonetheless indirectly related to the way we represent ourselves, our values, and societies. Bruce Sterling, writer both of science fiction and what he calls "design fiction", makes a strong point that contrary to science fiction, design fiction is submitted indirectly to user constraints. In an article on science-fiction, he mentions the indirect relations between literature and the "real world": "Many problems I once considered strictly literary are better understood as interaction-design issues. Literature has platforms. By this I mean the physical structures on which literature is conceived, designed, written, manufactured and distributed, remembered and forgotten. Literary infrastructure has user-experience constraints".[103] Lindley and Coulton,[104] elaborating on Bruce Sterling, also emphasize the role of diegetic probes, that is artifacts that support (and are supported by) a whole narrative introducing a new world: « the deliberate use of diegetic prototypes to suspend disbelief about change ».

I want to pause on this last quote because it introduces one last critical property of expansive literature. Lindley and Coulton's definition includes the "meta goal" of these productions. Designers using this type of probe, beyond staging new paradigms, also stage the possibility of change. Changes, options, variations, are the core demonstration of these stories. It is the revolutionary potential of design that is reaffirmed in the staging of diegetic probes and the use of expansive literature. They immerse the readers, audiences, and participants in the thrill of change. If we combine designers and engineers productions, this meta narrative about change is powerful because almost everything is taken care of: in addition to the artifact, we witness the gestures that go with it. The gestures are meaningful because they are embedded in a situation. The situation is coherent because all the elements converge to sustain the activities of the actors. The activities of the actors are relevant because

[103] Sterling (2009).

[104] Lindley and Coulton (2015).

we understand their values. Their values are authentic because they emerge from a long history made of political events, economic turmoil, etc. The strength of these convergent narratives is that they produce a whole world. The invention becomes part of a holistic view, not only a solution to a problem.

4.6 Conclusion: Design as Projection, Condensation and Expansion

This chapter has focused on two opposite poetic practices: condensation and expansion. Let's go back for a second to what this means.

First we have studied the semiotics of condensed forms of identity through names and logos. The main advantage of these practices is to explore the meaning of the invention, to develop as well as create new knowledge. The second advantage is to fuse the many faceted aspects of a technology, project, or research field in one word or image. Because they condense so many possibilities, these words and images also trigger a powerful process of interpretation that continues the process of invention while the object is being socialized. Condensation as a poetic practice is supported by an extremely difficult process of reduction, choice, redefinition. But it is also based on a potent metaphorical transformation since the words have to stand for a vast variety of "signified". Each element of the name (signifier) refers to different techniques, uses, values (signified). The logos put together images, texts, symbols whose original meanings are changed by their proximity and in contact with each other. I will further develop this metaphorical process in Chap. 5 where I look at design as composition in a field of tensions. Paradoxically, this extremely condensed form of poetic production can trigger a great amount of different interpretations based on the connotations. In the reception process, each person is free to redeploy, in her own way, the diversity of references behind the name or logo.

We have also seen that poetic practices include narratives such as use cases and scenarios that also build and expand the scope of research projects. The expansive properties of these literary productions are very different. They organize a whole world view which, on the one hand, narrows the interpretive options since it structures a precise view of the activity, but, on the other hand, offers a probable future with a great variety of details that all offer handles for memories (of other narratives, objects, situations) and projections of new applications and circumstances.

Engineers and designers therefore appear as poets and narrators who structure their invention not only through scientific means but through a dynamic meaning-making process. They are creators who compose with a variety of materials as I shall discuss in the following chapter.

References

Aristotle. (2005). *Poetics* (trans: Sachs, J.). Focus Publishing/R. Pullins Co.

Aristotle, & Lam, E. (2014). *Rhetoric* (trans: W. Rhys Roberts). Create Space Independent Publishing Platform.

Augé, M. (2009). *Non-places: An introduction to supermodernity*, trad. par John Howe, (2nd éd.). London: Verso.

Barthes, R. (1973). *Elements of semiology*, trad. par A. Lavers, & C. Smith. New York: Hill and Wang.

Bosredon, B. (1997). *Les titres de tableaux. Une pragmatique de l'identification*. Paris: PUF.

Bourriaud, N. (1999). *Formes de vie. L'art acont et l'invention de soi*. Paris: Editions Denoël.

Bucciarelli, L. L. (1988). An ethnographic perspective on engineering design. *Design Studies, 9*(3), 159–168.

Bucciarelli, L. L. (2002). Between thought and object in engineering design. *Design Studies, 23*(3), 219–231.

Burge, T. (1973, août 16). Reference and proper names. *The Journal of Philosophy 70*(14): 425. doi:https://doi.org/10.2307/2025107.

Burri, R. V., & Dumit, J. (2007). Social studies of scientific imaging and visualization. In E. J. Hackett (Ed.), *The new handbook of science and technology studies* (pp. 297–317). Cambridge, MA: MIT Press.

Butor, M. (1969). *Les mots dans la peinture*. Genève-Paris: Les sentiers de la aconteu éditions d'art Albert Skira/ Flammarion.

Cambrosio, A., Jacobi, D., & Keating, P. (1993). Ehrlich's "beautiful pictures" and the controversial beginnings of immunological imagery. *Isis, 84*(4), 662–699.

Chalmers, A. F. (1999). *What is this thing called science?* (3rd ed.). Indianapolis: Hackett Publishing.

Clausen, H. (1993). Narratives as tools for the system designer. *Design Studies, 14*(3), 283–298.

Clifford, J., Marcus, G. E., & Santa Fe, N. M. (Eds.). (1986). *Writing culture: The poetics and politics of ethnography : A school of American research advanced seminar*. Berkeley: University of California Press.

Cockburn, A. (2000). *Writing effective use cases* (1st ed.). Boston: Addison-Wesley Professional.

Crystal, D. (2005). *The stories of English*. London: Penguin.

Dickson, G. (1975). *Combat SF*. London: Doubleday.

Dourish, P. (2006). *Re-Space-ing Place: "Place" and "Space" Ten Years OnCSCW'06*. Banff, Alberta, Canada. Copyright 2006 ACM.

Eco, U. (1989). *The open work*. Cambridge, MA: Harvard University Press.

Genette, G. (2014). *Introduction à l'architexte*. Paris: Seuil.

Gentes, A. (2008). Design et aconteur aconteu dans les technologies de l'information et de la communication. *Hermès, 50*, 83–89.

Gentes, A. (2012). The design of emotions: How the digital is making us more emotional. In C. Pelachaud (Ed.), *Emotion-oriented systems* (1st ed.). New York: Wiley-ISTE.

Gentes, A. & Selker, T. (2013). Beyond rhetoric to poetics in IT invention, In: Paula Kotzé. et al. (Ed.), *Human-computer interaction – INTERACT 2013* (Lecture notes in computer science 8119, pp. 267–279). Berlin/Heidelberg: Springer. doi:https://doi.org/10.1007/978-3-642-40477-1_17.

Greimas, A. J. (1983). *Structural semantics: An attempt at a method*. Lincoln: University of Nebraska Press.

Hackett, E. J., & Society for Social Studies of Science. (2007). *The new handbook of science and technology studies*. Cambridge, MA: MIT Press. Published in cooperation with the Society for the Social Studies of Science.

Harrison, S., & Dourish, P. (1996). Re-Place-ing space: The roles of place and space in collaborative systems. In *Proceedings of the CSCW'96*. Boston: ACM Press.

Hebert, L. (2006). Tools for text and image analysis: An introduction to applied semiotics. Texto!, [online]. Available on: (Consulted 19 Sept 2017).

Henderson, P. W., & Cote, J. A. (1998, April). Guidelines for selecting or modifying logos. *Journal of Marketing 62*, 14–30. p. 14.

Huguet, F. (2016). *Recoudre avec du sans fil : Médiations technologiques et technologies de médiations*. Paris: Telecom ParisTech.

Jakobson, R. (1981). *Essais de linguistique générale*. Paris: les Éd. de Minuit..

Jameson, F.. (1982, juillet). Progress versus Utopia; Or, can we imagine the future?, *Science Fiction Studies* 9(2): 147–158.

Jameson, F. (2005). *Archaeologies of the future: The desire called utopia and other science fictions*. New York: Verso.

Jeanneret, Y. (1994). *Ecrire la science – Formes et enjeux de la aconteurion*. Paris: PUF.

Kapferer, J.-N. (1997). *Strategic brand management*. Great Britain: Kogan Page.

Kardash, J. (2008) Tech history: How bluetooth got its name, *EE-Times*, http://www.eetimes.com/document.asp?doc_id=1269737&page_number=1 (retrieved 19 September 2017).

Keller, K. (2012). *Strategic brand management: building, measuring, and managing brand equity* (4th ed.). Boston: Prentice Hall.

Kripke, S. A. (1991). *Naming and necessity* (New Edition). Oxford/Cambridge: Wiley-Blackwell.

Latour, B. (1988). *Science in action: How to follow scientists and engineers through society*. Cambridge, MA: Harvard University Press.

Lindley, J., & Coulton, P. (2015). Back to the future: 10 years of design fiction. In*Proceedings of the 2015 British HCI conference* (pp. 210–211). New York: ACM. http://dl.acm.org/citation.cfm?id=2783592.

Lloyd, P. (2000). Storytelling and the development of discourse in the engineering design process. *Design Studies, 21*(4), 357–373.

Lynch, M., & Woolgar, S. (1990). *Representation in scientific practice*. London: Mit Press.

McCarthy, J., Minsky, M. L., Rochester, N., & Shannon, C. E. (1955). *A proposal for the Dartmouth summer research project on artificial intelligence*. http://www-formal.stanford.edu/jmc/history/dartmouth/dartmouth.html

Meschonnic, H. (1970). *Pour la poétique, tome 1*. Paris: Gallimard.

Minsky, M. (2007). *The emotion machine: Commonsense thinking, artificial intelligence, and the future of the human mind*. New York: Simon & Schuster. Reprint.

Mitchell, W. J. T. (2002). Showing seeing: A critique of visual culture. *Journal of Visual Culture, 1*(2), 165–181. https://doi.org/10.1177/147041290200100202.

Musiani, F. (2012). *Nains sans géants: Architecture décentralisée et services Internet*. Paris: ENMP. http://www.theses.fr/2012ENMP0043.

Olohan, M. (Ed.). (2000). *Intercultural faultlines: Research models in translation studies: Textual and cognitive aspects v. 1*. Manchester: St Jerome Publishing.

Postman, N. (1993). *Technopoly: The surrender of culture to technology* (1st ed.). New York: Vintage.

Preminger, A., Brogan, T. V. F., & Warnke, F. J. (Eds.). (1993). *The New Princeton encyclopedia of poetry and poetics* (3rd ed.). Princeton: Princeton University Press.

Propp, V. (1968). *Morphology of the Folktale: Second Edition, Revised and Edited with Preface by Louis A. Wagner, Introduction by Alan Dundes*, éd. Par Louis A. Wagner, trad. Par Laurence Scott, (2nd éd). University of Texas Press. p. 25.

Rancière, J. (2006). *Le Partage du sensible : Esthétique et politique* (La Fabrique, 2000). Jacques Ranciere et Gabriel Rockhill, *The politics of aesthetics*, (1re éd.), Continuum.

Rastier, F. (2001). *Arts et sciences du texte*. Paris: Presses universitaires de France.

Reeves, S. (2012). Envisioning ubiquitous computing. In*Proceedings of the SIGCHI conference on human factors in computing systems* (pp. 1573–1582). New York: ACM. http://dl.acm.org/citation.cfm?id=2208278.

Ricoeur, P. (1990). *Time and narrative, Volume 1*, trad. par K. McLaughlin & D. Pellauer, reprint, Chicago: University of Chicago Press.

Rutledge, J. D., & Selker, T. (1990). Force-to-motion functions for pointing. In*Proceedings of the IFIP TC13 third interational conference on human-computer interaction, INTERACT '90* (pp. 701–706). Amsterdam: North-Holland Publishing. http://dl.acm.org/citation.cfm?id=647402.725310.

Sack, W. (2013). Une machine à aconteur des histoires: Propp et les software studies (Transl. Sophie Bargues Rollins), Revue les temps modernes, 5/676. Paris: Gallimard, pp. 216–243.

Searle, J. R. (1958). Proper names. *Mind, 67*, 166–173. http://mind.ucsd.edu/syllabi/00-01/phil_lang/readings/searle-01.html

Sicard, M. (1995). *Chercheurs ou artistes ? Entre art et science, il rêvent le monde*. Paris: Autrement.

Simondon, G., & Chateau, J.-Y. (2005). *L'invention dans les techniques: Cours et conférences*. Paris: Seuil.

Star, S. L., & Griesemer, J. R. (1989). Institutional ecology, 'translation' and boundary objects: Amateurs and professionals in Bereley's Museum of Vertebrate Zoology, 1907–39. In*Social Studies of Science* (Vol. 19). London: Sage.

Sterling, B. (2009). Design fiction. *Interactions, 16*(3), 20–24.

Turner, P., & Turner, S. (2003a). Two phenomenological studies of place. *People and computers XVII – proceedings HCI conference*, pp. 21–35.

Turner, S., & Turner, P. (2003b). Telling tales: understanding the role of narrative in the design of taxonomic software. *Design Studies, 24*(6), 537–547.

Valéry, P. (1957). *Oeuvres, tome 1*. Paris: Gallimard.

Chapter 5
Design as Composition of Tensions

5.1 How to Organize a Design Crisis?

The irruption of radical designs creates crises that trigger the actors to dramatically question the status quo, reorganize themselves, introduce new products or new services, new values, and new aesthetics. Radical design challenges our habits, our views of the world, our comfortable set of ideas, and the organization of power. Eventually, these metamorphoses beget new configurations that we learn to understand and appreciate. The history of art and literature is the history of the permanent outrage brought by the new forms that pit the Moderns against the Ancients.[1] Each aesthetic new age has brought intellectual and public outcry where the merits of imitating the former generation of creators were debated. Whether in the seventeenth century with Charles Perrault[2] praising the "new way" against Boileau[3] who advocated the Antiquity as a model, or two centuries later, with Victor Hugo's play: Hernani (1830) that heralded the Romantic area, the clash between the protagonists is more often than not violent (Hernani is better known today for the intellectual – sometimes physical – battle that it triggered rather than for the play on its own merits). I think that we should keep in mind the fact that every design project is a quarrel of the Ancients and the Moderns. Whether art has more latitude to explore entirely new aesthetics, and design practice has to juggle between radical invention and easing itself into people's lives is beside the point. The crisis is embedded within the creative process because researchers as well as artists and designers have to extricate themselves from the seemingly pre-determined configuration of things.

---

[1] Armogathe et al. (2001).

[2] Charles Perrault (1628–1703) was a French writer, who, amongst other things wrote the famous Tales of Mother Goose that introduce the fairy tale genre. He led the "Modernist" movement in the seventeenth century, praising the new ways of writing that fitted best the King's times (Louis the XIV).

[3] Nicolas Boileau-Despréaux (1636–1711) was a French poet and critic who endeavoured to define the rules of poetry. He led the faction of the Ancient in praising the art of the poets of Antiquity.

A. Gentes, *The In-Discipline of Design*, Design Research Foundations,
DOI 10.1007/978-3-319-65984-8_5

The question we need to consider is how they achieve such a revolution but also how, after climax and anti-climax, the new situation is accepted or at the very least understood as coherent. After all, new genres do not bring chaos forever.

To begin with, a state of permanent crisis defines the design work itself as a process and as a result. To answer one question, designers produce multiple scenarios, with different interactions, using different materials and techniques. In my experience with engineering and design students, the process is stressful not only because it taxes everybody's creativity but also because it is totally counterintuitive to the idea that there is one and only one good answer to a problem. It is also stressful because we ask the participants to accept and work on controversies more than they usually do. It is not only a confrontation of ideas but of media, textures, technologies, tools. Design education and design practices are repeatedly bringing in methods of confrontation or at the very least methods that diversify the sources and tools of design. They organize what I suggest to call "a field of tensions" where the main skill is not so much to manage a series of steps but to "compose" with different elements and properties to unfix views and challenge knowledge, and eventually to generate new coherent situations.

Rather than focusing on the psychological or managerial aspects of the question, I want to share observations about situations that deliberately orchestrate this field of tensions between materials, medias, and tools. The chapter is thus called "Design as composition of tensions" for three reasons. First, it is a pragmatic observation of how designers gather material to do their job. I like to show my students what an artist's studio looks like in particular Bacon's, Calder's, or Pollock's.[4] They are full of stuff, cluttered with artifacts, tools, cultural products, textures and materials that are arranged, rearranged, organized and disorganized. While collections are rationalized through archives like the "materiautheque"[5], artists' studios look like curiosity cabinets as David Hockney presented in his "Great Wall".[6] Second, it is a reflection on methodologies. The word "tension" is used here to describe the fact that heterogeneous elements are brought together to un-fix, that is to go beyond preconceptions and expand the design space. Third, to look at design as composition of tensions is to make a hypothesis regarding its epistemology. I will argue that "projective abductive processes" organize the whole composition activity. Abduction proper is a semiotic practice that brings to the forefront unforeseen connections out of a diversity of elements. What I call "projective abduction" is a semiotic practice that builds a world to be. Hence, tensions are solved in the new composition.

I will, therefore, try to understand why a field of tensions is both a disruptive force and a way of looking at the reorganization of knowledge in a new aesthetic. Doing so means switching from a temporal, chronological, model of the design project to a spatial, topological, model of design and to look at how the expression "design space" can be interpreted as a "matrix", as the French philosopher of design

[4] Some pictures of famous artists' studios can be found on this website: http://www.artistsandillustrators.co.uk/news/Buildings-Architecture/530/famous-artists-studios

[5] See for instance, in Paris, http://www.lelieududesign.com/la-materiautheque-materio

[6] Hockney (2006).

Pierre-Damien Huyghe, following Paul Klee's theory, suggests. To describe the designer's stage and elaborating on Huyghe's definition, I am trying to understand the design clutter as an apparatus that bridges different knowledge bases, power stakes, and aesthetics, so as to produce a new composition. Again, the history of art shows a great deal of possible compositions that each time defy what was expected in terms of aesthetics. At the same time, these unexpected compositions were ways to produce new meanings. I think that we need to take a closer look at how engineering research and design also compose.

To support my hypothesis, two examples will be presented in this chapter. I do not pretend that they are the only ways for researchers and designers to compose, but they came as a surprise to me, even as a quite unsettling experience when they challenged my personal relationship to writing tools. More importantly, they focus on material and media on the one hand, and on tools on the other, which is a way to tackle these two important nonhuman "actors" in the design space, while I deal with human stakeholders in Chap. 6. The first example presents a project where designers assembled a diversified corpus of heterogeneous sources. It took place in 2012 when the Codesign Lab helped design an innovative e-learning platform. We followed a group of designers and researchers to see how semiotic knowledge was acquired and then transferred to a new product within what we finally described with Marie Cambone as a "contradictory semiotic analysis" or a "contrasting semiotic analysis" to describe a confrontation of different sources for design. The second example focuses on "tools". It took place in 2002 and involved a multimedia artist, a group of students, and three professors who worked together to write an interactive show. The participants used different tools for the same purpose. On the one hand, they played and were played by these tools, which channeled their writing skills. On the other hand, using multiple writing tools was also a way to contrast different interpretations and to expand the scope of the design work. Thanks to Mathias Bejean, we came up with the word "constellation" as the best way to describe not a linear process but a group of versions of a theatrical play, all valid in their own ways. Both examples show design practices that organize the confrontation of design elements, and that play on tensions inherent in bringing diverse material together.

5.2 "Contrasting Semiotic Analysis": The Semiotic Organization of a Confrontation[7]

Experienced designers work with a wide-range of artifacts and media, technologies, contents and visual representations. This wide-ranging experience makes it possible for them to come up with an intricate balance of known and unknown in

[7] First versions of this section were published in Gentès, Annie, Cambone, Marie, « Designing empathy: the role of a "control room" in an e-learning environment », Journal of Interactive Technology and Smart Education, 2013.

the form of unexpected configurations of signs and forms. Closely following their work means seeing how they tap into their experience to build new designs. Books that teach design or architecture emphasize the need to learn through studying former buildings or artifacts.[8,9] The role of former knowledge in design is therefore extremely important and has been studied, for instance, by Willemin Visser who points out the reuse of knowledge in different circumstances and fields of design:

> Reuse of knowledge (from specific previous design projects) through analogical reasoning has been observed in many cognitive design studies as a central approach in design."[10] Former knowledge plays a part in the heuristics of the project as design memory[11] or design precedents.[12]

Eilouti also looks at design precedents and how they are part of an analogy process that feeds new artifacts. Closer to genre theory, the author shows that typologies are also a way to gather similar elements thanks to the identification of certain of their properties that can therefore be re-used and combined at this more abstract level.

> Typology can be described as the enumeration and categorization of collections of components based on pre-defined criteria in order to reflect certain characteristics of the individual components and relations among them in their combinations.[13]

Finally, understanding the common underlying structure of particular artifacts serves as a starting point for design/practice as well.

In other words, designers have to start somewhere. Analogy with previous artifacts and situations, typology of interactions and forms, and the analysis of former compositions, help at several stages of the design process to suggest ideas, to implement patterns in prototypes, to evaluate the design of the object. In this respect, the question of fixation seems particularly acute: though it makes perfect sense that a designer needs to learn from former projects and objects, how does he/she avoid getting stuck in a particular example, and repeating the same patterns? This part of the chapter considers how semiotic analyses play a central part in using design precedents. In the first case, there was an intense analytical and comparative stage that was instrumental in discovering structural and semiotic characteristics of a genre of artifacts as well as in displacing certain features and principles that were reinterpreted and embedded in a new configuration. This "contrasting semiotic analysis" will be detailed in the following section.

---

[8] Leupen et al. (1997).

[9] Unwin (2009).

[10] Wisser (2006).

[11] Oxman (1994).

[12] Eilouti (2009).

[13] Eilouti (2009).

Fig. 5.1 3S Informatique's visualization of the project VUE. 2010

5.2.1 *First Case: The e-Learning Platform VUE*

VUE[14] (Fig. 5.1) is a research project on multimodality that was developed in a partnership between a service development company (Groupe 3S Informatique), the Signal and Image processing Department of Telecom ParisTech with Jean-Claude Moissinac, and the Codesign Lab. A team, including researchers in computer science, design, information and communication sciences, worked one year to develop an e-learning platform that prototyped specific ways of storing data so that it could be adapted and used on various media (computer, tablet, smartphone).

The Codesign Lab was in charge of several tasks: a survey of distant teaching and e-learning services; the definition of a set of specifications to develop the technical platform; the design of graphical interfaces; and, the definition, organization, and analysis of the end-user tests.

The project took place before the worldwide progress of MOOCs (what certain journalists called the tsunami MOOC in 2012). Nonetheless, in 2010, there were many e-learning platforms that provided similar services. As it were, a number of analyses were already available and all showed that attendance in distant learning was a recurrent issue. To encourage continued participation, educators primarily focused on designing activities promoting collaboration and interaction between students. In a virtual context, it requires the implementation of technical and social mediations because the distance is not only physical but also technical, socio-cultural, socio-economical and educational.[15,16,17] The interplay of learning and technology that was studied in particular by CSCW (computer-supported cooperative work) researchers, especially CSCL (computer-supported collaborative

[14] VUE means "sight" in French.

[15] Moore and Kearsley (2011).

[16] Holmberg (1995).

[17] Jacquinot (1993).

learning),[18] also showed that the way participants were represented influenced interaction.[19,20] On the basis of this first survey, the team decided to focus on the problem of attendance. However, the analysis of the situation was not enough to come up with an innovative proposal. The ethnography of long distance learning did not give us formal design cues. The team therefore turned to a semiotic analysis of e-learning platforms.

5.2.2 "The Loneliness of the Long-Distance Learner"

The design team observed that a virtual platform is not only a space for coordination between actual people, it is also a virtual platform that turns participants into actors and characters who play a role in a space that they animate. Building on the metaphor of the computer as a theater,[21] they considered that the screen was a dramaturgic space where represented interactions were taking place. After assembling a corpus of e-learning websites, the research team looked for recurring features in e-learning platforms so as to gather structural properties, functional and aesthetic qualities that defined the e-learning "genre". They also researched and analyzed the various representations of presence.

As we saw in Chap. 4, a corpus is a group of texts (linguistic and/or visual) or artifacts that is used implicitly in everyday life as a way to organize knowledge and communication.[22] But, in a research setting, corpuses are gathered by the researcher to analyze their characteristics and to validate or not their commonality so as to answer a research question. Often the question is to know if they belong to the same genre or if, on the contrary, some of the examples are original, either radically or moderately.[23] In any event, the corpus is not only formal but also pragmatic. The elements show what Wittgenstein called a "family resemblance" that is objects that are similar not because they share the exact same formal features but because they are considered to be analogous on a certain level by social actors. As Lakoff further developed: "interactional properties are prominent among the kinds of properties that count in determining sufficient family resemblance".[24] Taking the example of chairs, he observes that:

> The interactional properties relevant to our comprehension of chairs will include perceptual properties (the way they look, feel, etc.), functional properties (allowing us to sit), motor-activity properties (what we do with our bodies in getting in and out of them and while we're in them), and purposive properties (relaxing, eating, writing letters, etc.).

[18] Stahl et al. (2006).

[19] Stahl et al. (2006).

[20] Blandin (2004).

[21] Laurel (1993).

[22] Lakoff et Johnson (1980).

[23] Rastier (2002).

[24] Lakoff et Johnson (1980).

In our case, the team selected four e-learning platforms that had been identified as targeting the same audience (people involved in continuing education) and that offered similar services. We conducted a semiotic analysis on the following points: rendering of environments, representation of actors, terms of interaction, and possibilities of changing points of view (Figs. 5.2–5.4).

We then realized that most platforms strive to strike a balance between representing the students and offering working tools. Two aspects seemed equally important. First, users are given some leeway to personalize their tools and working environments. Second, users are given the possibility to access two screen spaces: the virtual rendering of the class (either a table of webcams or a 3D environment – Second-life like — not represented here) and the toolbox. Even so, as can be seen on the screenshots above, no interface represented the classroom as a whole with a compelling orienting view. What is more, students were not always shown on the screen (Fig. 5.2: only the teacher is present). On other interfaces, students were present through a line of their webcams (Fig. 5.3) or a table of their webcams (Fig. 5.4). These design choices made it very difficult to represent all students on the screen and thus did not easily contribute to a feeling of belonging to a class. In addition, these four platforms offered a single type of class: the conference mode. Group work or tutorials were not taken into consideration.

Then, the team was convinced that the "loneliness of a long-distance learner" was a major challenge. However, this motivational and psychological metaphor had to give way to a design solution that would actually offer a service through an interface. The question became how to fight the feeling of separation from the group, the loneliness, and provide a feeling of togetherness not only through activities but also from the interface and the representation of the situation. The combined questions

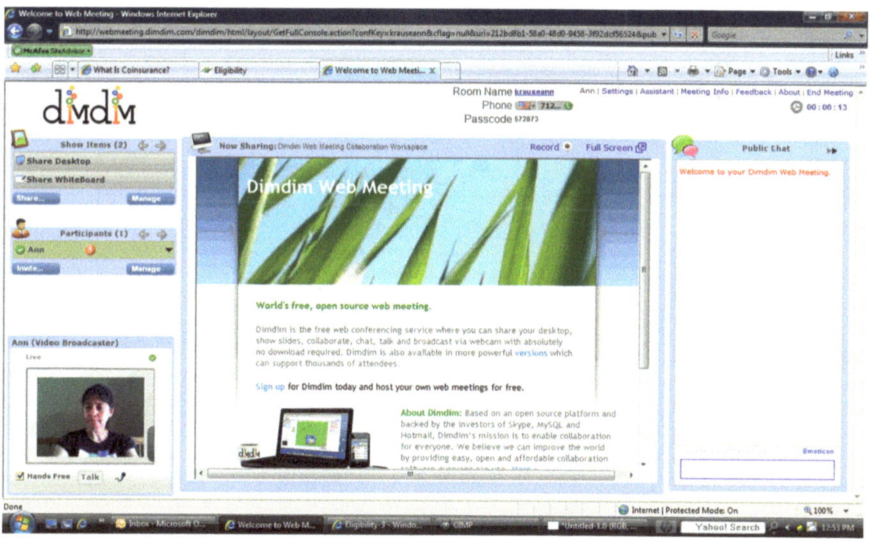

Fig. 5.2 Dim Dim e-learning platform (http://www.dimdim.com/)

Fig. 5.3 ISL iMeeting (http://v5.islonline.com/isl-groop/overview.htm)

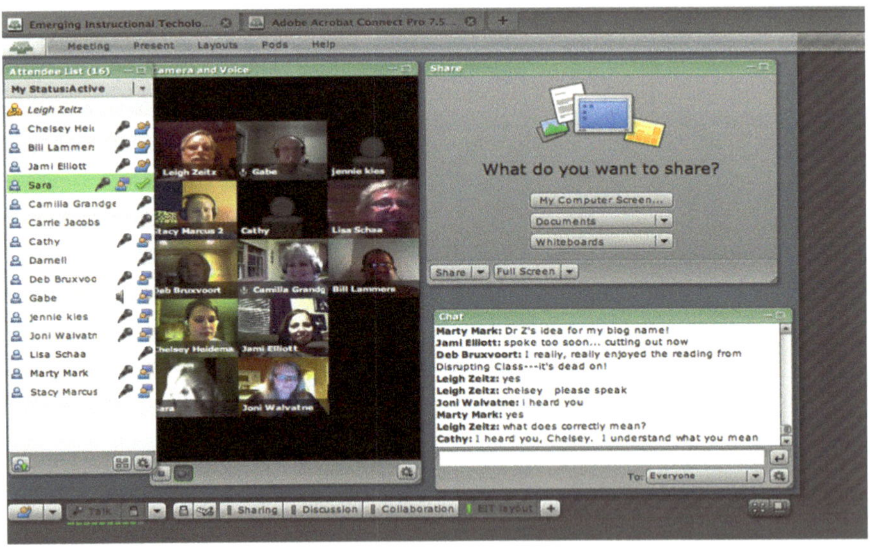

Fig. 5.4 Adobe connect (http://www.adobe.com/fr/products/connect/)

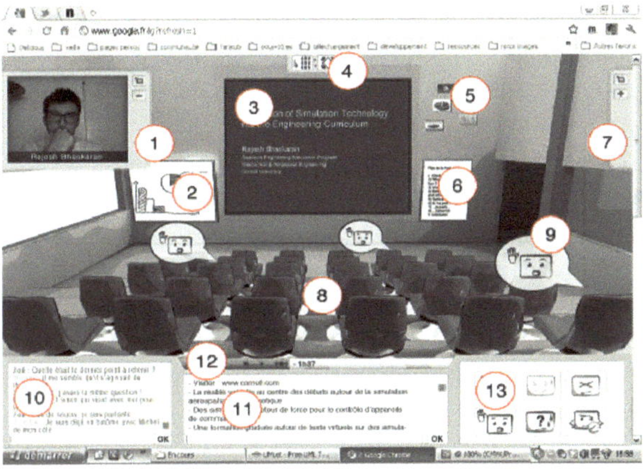

1. Teacher Webcam
2. Whiteboard
3. Slides
4. Permutation « lecture » interface to « group work » interface
5. Documents
6. Lesson plan
7. student Webcam
8. Classroom
9. Smileys to represent a student's question
10. Chat
11. Note-taking
12. Timeline
13. Smileys

Fig. 5.5 Description of the interface for the student

of representation of self and others, global view of the group, and tools to communicate in multiple ways, led the team to emphasize points of view and changes of perspective, in particular because the previous interfaces did not connect different points of view to different activities. The keyword here is "viewpoint". If one switches from the psychological to the visual plane, the situation can be described as managing different viewpoints during the interaction. The points of view are given by the actual position of the interlocutors. A person can, therefore, imagine the reverse shot of what she sees. Also important was the fact that the "togetherness" of a real class was not represented on the screen and that we needed to find a point of view that could represent it.

At that stage, the designers left the design space of e-learning platforms. They felt the need to explore how these feelings (togetherness) and representation of self and others were dealt with in other media to try to create an equivalent in VUE, a new branch in the family resemblance tree.

5.2.3 "Empathy" and Togetherness in Other Media

After the first analysis of e-learning platforms, the designers delineated another corpus composed of visual media: painting, photography, film, and comics. They especially analyzed how each media aesthetically treats the issue of self-representation and the representation of different protagonists at the same time in different spaces (to see different scenes at the same time, to make ellipses in space, etc.) So, the design team analyzed how these different media dealt with the issue/concept of viewpoints.

In film and video, they noted that the viewer shares the viewpoint of the characters either through a first person narrative viewpoint or with medium shot (to feel closer to the actors, to simulate an immersion in the scene) and close-ups (to show facial emotions).[25,26] But empathy is not only built through first person narrative. Other shots give the spectator a feeling that she is part of a group, that she can share the general view that characters have from inside the movie: the experience of dialog in a group can be rendered through bird eye view or ¾ shot that have all the actors visible to the camera (in particular to allow complicated dialog scenes between more than two people without changing camera position). Editing techniques in cinema (and subsequently in video and 3D environment) provide dynamic change of viewpoints: shot and reverse-shot for example, to simulate a dialog, etc.[27,28]

The design challenge was to emulate both the visualization of different viewpoints and to provide the users with a flexible capacity to choose their viewpoint as they wished and in relation to their activities. The main question was: how does an interface support this process when trying to represent a group of 15 to 30 learners while also creating a way to represent team work (4 students) on an interface?

5.2.4 *The Result: VUE as a Digital Control Room to Fight Loneliness*

Alone behind their computers, students easily drop out of class. To fight the loneliness, the team decided to focus on supporting a community through several video options.[29] In conference mode, the designers produced two interfaces: the student's interface (Fig. 5.5) and the teacher's interface (Fig. 5.6). The interface of the teacher is a reverse-shot of that of the student. While VUE used a number of elements and tools that were observed in other e-learning platforms,[30] it also came up with original features such as the large place occupied by the classroom. The teacher faces the class as in a real situation. But more importantly, VUE supports access to multiple points of view of the class thanks to a control room.[31] The "digital control room" of VUE enables the user to select the "camera" that suits her need to understand the situation, to participate more effectively, or on the contrary to create some distance.

[25] Aumont et al. (1992).

[26] Doane (2003).

[27] Aumont et Marie (2004).

[28] Oudard (1969).

[29] A more detailed description of VUE can be found in Gentès and Cambone (2013).

[30] They included a space for slide presentations (which often occupies most of the screen), live performances of the teacher, student representation (via webcam, photos, avatars, virtual agents, a list of names...), communication tools (chat) and sometimes elements to measure the mood of the class (smileys, color code ...).

[31] The control room in television broadcast is the place where the video feeds from the different cameras can be watched. The production team selects the video feed that is going to be broadcasted by TV channels.

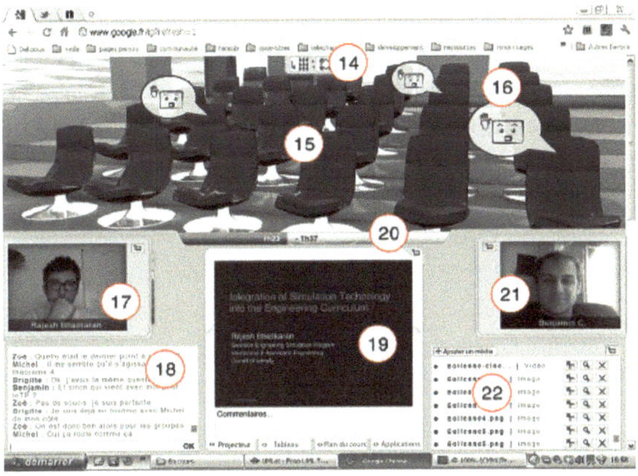

14. Permutation « lecture » interface to « group work" interface
15. Classroom
16. Smiley to represent a question of a student
17. Webcam teacher
18. Chat
19. Slides
20. Timeline
21. Webcam student
22. Other medias used during the lesson

Fig. 5.6 Description of the interface "teacher"

Fig. 5.7 Codedoc project, Whitney Artport, 2002. http://artport.whitney.org/commissions/codedoc/ (retrieved 20 September 2017)

It is a visual tool directly affecting the way the information is accessed. Each user becomes his/her own digital screen director.

5.2.5 Definition of the "Contrasting Semiotic Analysis"

If we sum up the different planes of composition, we see that the survey raised a question: how to help with the attendance issue. But the question was not enough to come up with design ideas. The team therefore did two semiotic analyses. First, their analyses of e-learning platforms gave them the recurring features of the systems. Second, their analysis of viewpoints in different media provided a syntax to deploy so as to enable users to change and share viewpoints. The team deployed a "contrasting semiotic analysis", an expression we coined with Marie Cambone to address this back and forth "dialog" between two corpuses and semiotic analyses that balanced the convergent effects within a genre (learning websites or points of view in cinema) and the diverging effects of bringing together two different corpuses where each element is contrasted so that new associations can arise. The first analysis looked at general features of the e-learning genre and reinforced the commonality of the genre and the second one was a direct counter-proposition on the basis of semiotic features borrowed from other media. The main advantage of using representations and media as a starting point was that it prevented any attempts to emulate "real life". The double and contrasting semiotic analysis helped the team to

- focus on the visual interface itself as a stage rather than to consider it as a mere tool for e-learning and therefore opening the opportunity to play with the semiotic possibilities of the system.
- rethink the question of users' representation and find an equivalent to a close-up in cinematic grammar,
- deal with the complicated issue of loneliness by allowing participants to change and share viewpoints.

More generally, the exploration of media based representations helped craft alternatives to support experiences that cannot be lived in "real life": to see different scenes at the same time; to be both very subjective and omniscient; to make ellipses in time and space.

As mentioned earlier, these observations fit what Schön calls the interaction between the designer and her material. He shows how a designer "shapes the situation, in accordance with his initial appreciation of it, the situation "talks back" and he responds to the situation's back talk".[32] But a situation does not talk back if a system of tensions and confrontations between social, technical, and semiotic systems is not organized as such. In other words, it seems important to define a meta-communication system that specifically addresses this question of confrontation. Here the "contrasting semiotic analysis" seems precisely to be one of the meta-communication systems since it turns a survey of existing platforms into a confrontation of services and contents on different media. It is therefore part of the converging / diverging semiotic process that expand concepts (here sharing / changing viewpoints) so that they can become operational in mediated interactions.

The next section presents another example of a meta-communication system that brings the differences to the forefront, while considering the different results as part of the same design continuum.

[32] Schon (1984).

5.3 Using Several Tools as a Confrontational Technique[33]

If we only focus on the results, we neglect the role of artifacts, machines, tools that creators handle and that shape the project too. Certain artists, for instance those who participated in the digital art exhibition CODEDOC,[34] make it clear that code is both their material and tool since certain types of code bring certain types of artwork (Fig. 5.7).

> *CODeDOC* takes a reverse look at 'software art' projects by focusing on and comparing the 'back end' of the code that drives the artwork's 'front end'– the result of the code, be it visuals or a more abstract communication process.

Hence some harsh criticism of certain software that do not seem to support creative activities but incorporate inane bureaucratic practices and legitimize institutional writing. I am thinking of Edward Tufte's criticism[35] of Powerpoint. Hence too the promotion of self-made software by some researchers like John Maeda[36] or Alan Kay.[37] These designers warn that the user can fall prey to the underlying model of their writing tool and I have to admit that it is exactly what happened to me and a group of professors and students in the following experience. Our experiment with a multimedia artist demonstrated how tools could shape our writing, how we actually were writing under influence!

Here, I want to show two things: first, software are somehow inhabited not only by a figure of the "user"[38] but also by a figure of the "text". Software – more or less explicitly – have a definition of what a text is, what information is, and what it is to read and write. Second, a tool is not only an artifact but also an apparatus and therefore can structure the way people contribute or create.

5.3.1 A Lesson from Art: Designing a Three-Stage Show

In 2002, with two researchers from the Department of Computer science, Alain Grumbach and Jean-Claude Moissinac, we invited twelve engineering students to participate in the early design phases of a live, interactive performance, to be held simultaneously on three separate locations connected with a (VTHD[39]) broadband network. My school wanted to test its broadband network but also wanted to explore the aesthetic and social potential of connecting three spaces that were both virtual

---

[33] First versions of this section were presented at the IASDR conference Gentès and Béjean (2011).

[34] http://artport.whitney.org/commissions/codedoc/index.shtml."

[35] Tufte (2006).

[36] http://www.maedastudio.com/index.php

[37] Alan Kay, "People who are really serious about software should make their own hardware," http://en.wikiquote.org/wiki/Alan_Kay

[38] Akrich (1990).

[39] VTHD: Vraiment Très Haut Débit.

Fig. 5.8 Mephisto circus. Art Project at Telecom Paristech, 2002–2003

and real. The group was invited to imagine a scenario based on Goethe's play "Faust" (1829–1832), hence the name of the show: "Mephisto Circus" (Fig. 5.8).

We worked with the French video and multimedia artist Michel Jaffrennou and with the company Gaia and its director Guilhem Pratz. Originally a video artist, Michel Jaffrennou had been working for quite a while on the mix between real and virtual spaces for such shows as "Le Petit Théâtre de Diguiden" (Diguiden's Small Theater).[40] The creative sessions had to provide a scenario of use and indications for the technical requirements and feasibility. It should be noted that "Mephisto Circus" was eventually not produced for a number of reasons, including the cost of such an operation and the difficulty of adapting the available spaces (conference rooms or class rooms of the connected institutions in Sophia-Antipolis – close to Nice – Paris, and Brest (Fig. 5.9) to the needs of the theater (backstage, complete darkness, etc.) Nevertheless, the artist incorporated some of these ideas, for example in "The Phantom Public", 2005, produced with sound designer Thierry Coduys, whereby the public could vary the lighting and sound of the show at whim.[41]

---

[40] de Meredieu (2005).

[41] Latour et Weibel (2005).

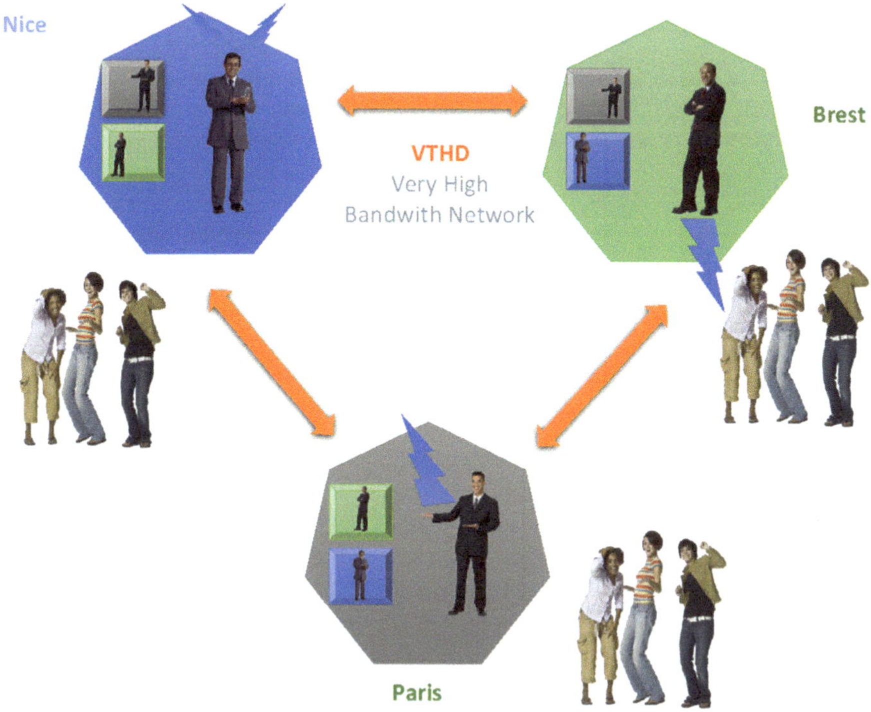

Fig. 5.9 Mephisto. Technical representation

5.3.2 From Writing a Text to Exploring Writing Tools

Before initiating the design work, all the participants of the group had to read Goethe's Faust that was the inspiration for the show. Then, brainstorming and writing sessions were organized with Michel Jaffrennou. During these sessions, the group produced a first text that was the starting point for the design process. To do so, the group used "Microsoft Word" as the more "natural" tool to tell a story. It allowed them to describe the characters as well as create the dialogs. For instance, they briefly defined the devil who would be the main character. This use of Word comforted their assumption that a show was a sequence of dialogs, as it is mostly taught in the French system of education. They thought that "designing a show" was like "writing a series of dialogs" that would then be staged with props, costumes, etc. In fact, stuck with a traditional model of theater, the participants did not imagine for a second that the public could be participative even though the goal was explicitly to create an interactive show. The use of the software Word did nothing to contradict this vision which we came to realize was not Jaffrennou's.

Coming from a contemporary branch of theater that combines different traditions (e.g. Antonin Artaud[42]) and very much influenced by the circus tradition,

[42] Artaud (1994).

Jaffrennou emphasized the physical impact of theater and its magical inspiration as well as the participation of the public. Without openly criticizing the results of the first writing sessions, Jaffrennou suggested that you use a flow chart (see Fig. 5.1), which enhanced the temporal structure of the show and its optional features triggered by the spectators. For instance, if the audience lighted the stage, Mephisto then disappeared. If the audience left the stage in semidarkness, Mephisto continued to play. When the group finally understood that the spectators were going to be part of the story, it radically changed the way they looked at the status of the audience: suddenly they were not only spectators but they became Faust, disillusioned and manipulative, trying some stage magic. They could be tricked by the devil, but they were also given some real power that could destabilize either Lucifer or another group of spectators (Fig. 5.10).

But this representation was not very helpful to visualize the three stages together and their interconnections. The artist suggested writing the show in "html pages" that would contain actors and actions and show the branching plot (Fig. 5.11). The hypermedia links between the different elements slowly turned the linear plot into the architecture of three interconnected places. The group was in fact struggling to move from a 2D representation to a 3D representation (Fig. 5.13).

Again, the artist shifted the group's emphasis from linear plot to a wider consideration of what is on stage and how the spectators can act in the play. We discussed the spectators' interventions (the why) but we did not really think about how: what would be the concrete props that could create not only a meaningful relationship between the stage and the actors but that could actually impact the progress of the show? The use of a spreadsheet application (Microsoft Excel), (see Fig. 5.12) took the group beyond the notion of the stage as a symbolic environment, to focus on the place as an ensemble of technical equipment and people. Jaffrennou had transformed this tool into a list "a la Prévert"[43] made of possible landscapes, characters, objects, etc. that could be diversely combined so that ideas of scenarios would appear by association. The show was then written again with a spreadsheet, introducing multiple objects and places as potential actors of the show (Fig. 5.12).

It should be noted that Michel Jaffrennou worked with the sociologist Bruno Latour with whom he shares an interest for the "grid" format as a flexible tool to play with signs. In particular, Jaffrennou used Latour's concept of "pedocomparator" defined as follows: "in *the regularity of its cube, their disposition in columns and rows, their discrete character, and the possibility of freely substituting one column for another, the pedocomparator belongs to sign. Or rather, it is through the cunning invention of this hybrid that the world of things may become a sign*".[44]

To consider the spatial dimension of the show, Jaffrennou finally led the group to use Adobe Director, an application that composes multimedia presentations (see Fig. 5.13). We could include the different tangible artifacts within a 3D representation with some perspective and the localization of screens and actors. The architec-

[43] Jacques Prévert (1900–1977) was a French poet and screenwriter who introduced the list as poetic material in the poem "Inventory" ("Inventaire" in the collection of Poems: Paroles, 1946).

[44] Latour (1999).

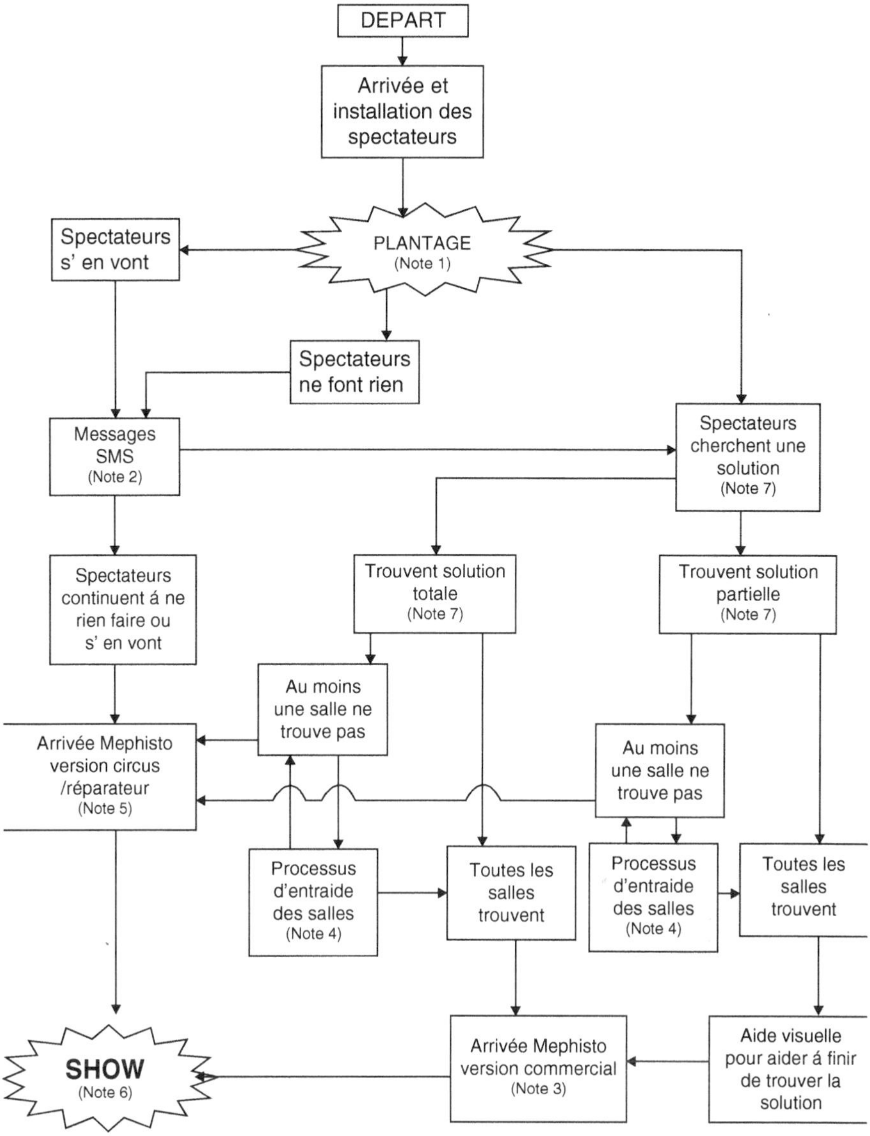

Fig. 5.10 Mephisto circus: Flowchart

tural dimension of the show and its scenography was at last fully perceived by the group. We no longer thought about the location as décor. Through Director, the value of artifacts not only as signs but as triggers to activate a new state of the show, was discovered. Eventually, we could define the show as a global architecture of events, people, and loci (the physical and symbolic spaces proper) (Fig. 5.13).

HARMONIE

Description brève de la fiche

Paris	Brest	Nice

Totems	*?*
Ecrans	*Actifs.*
Acteur	*Absent.*
Lumières	*ambiance lumineuse*
Son	*ambiance sonore*
Eléments utilisés	*liste des éléments déjà utilisés*
Action	*Les spectateurs utilisent les technologies de manière harmonieuse. Les différentes salles ont choisi de collaborer pour que le spectacle reste vivable. Mephisto est vaincu, l'élaboration d'un document collectif entre les salles vient sceller cette union qui permet de rejeter Mephisto.*

Description de la suite 1	*Description de la suite 2*	*Description de la suite 3*
Suite 1	Suite 2	Suite 3

Fig. 5.11 Mephisto circus – webpage

	A	C	D	E	F	G
1		ORINATEUR PILOTE				
2	TYPOLOGIE	SCENARIO	EVALUATION	TRAITEMENT OUT	QUI/QUOI ?	minute 001
3	HUMAIN/ACTEUR	scénario/acteur-1		ordres ecrits	acteur1	prompteur1
4	HUMAIN/ACTEUR	scénario/acteur-2		ordres ecrits	acteur2	prompteur2
5	HUMAIN/ACTEUR	scénario/acteur-3		ordres ecrits	acteur3	prompteur3
6						
7	HUMAIN/TECH/SPECT/PHONE	scénario/spect-1			téléphon	
8	HUMAIN/TECH/SPECT/PHONE	scénario/spect-2			téléphon	
9	HUMAIN/TECH/SPECT/PHONE	scénario/spect-3			téléphon	
10	HUMAIN/TECH/SPECT/PHONE	scénario/spect-4			téléphon	
11	HUMAIN/TECH/SPECT/PHONE	scénario/spect-5			téléphon	
12	HUMAIN/TECH/SPECT/PHONE	scénario/spect-6			téléphon	
13	HUMAIN/TECH/SPECT/PHONE	scénario/spect-7			téléphon	
14	HUMAIN/TECH/SPECT/PHONE	scénario/spect-8			téléphon	
15	HUMAIN/TECH/SPECT/PHONE	scénario/spect-9			téléphon	
16						
17	HUMAIN/TECH/SPECT/OBJET	scénario/objet-1			objet1	
18	HUMAIN/TECH/SPECT/OBJET	scénario/objet-2			objet1	
19	HUMAIN/TECH/SPECT/OBJET	scénario/objet-3			objet1	
20	HUMAIN/TECH/SPECT/OBJET	scénario/objet-4			objet1	
21	HUMAIN/TECH/SPECT/OBJET	scénario/objet-5			objet1	
22	HUMAIN/TECH/SPECT/OBJET	scénario/objet-6			objet2	
23	HUMAIN/TECH/SPECT/OBJET	scénario/objet-7			objet2	
24	HUMAIN/TECH/SPECT/OBJET	scénario/objet-8			objet2	
25	HUMAIN/TECH/SPECT/OBJET	scénario/objet-9	temps/vitesse 1		objet3	
26	HUMAIN/TECH/SPECT/OBJET	scénario/objet-10	temps/vitesse 2		objet3	
27	HUMAIN/TECH/SPECT/OBJET	scénario/objet-11	temps/vitesse 3		objet3	
28	HUMAIN/TECH/SPECT/OBJET	scénario/objet-12			objet3	
29	HUMAIN/TECH/SPECT/OBJET	scénario/objet-13			objet3	
30						

Fig. 5.12 Mephisto circus – excel file

5.3.3 Mixing Software: The Organization of the Confrontation

The observation shows that the participants' writing was influenced by the software because they engage a vision of what a "proper theatrical text" is: either a series of dialogs or a series of events, either the architecture of living and non-living actors

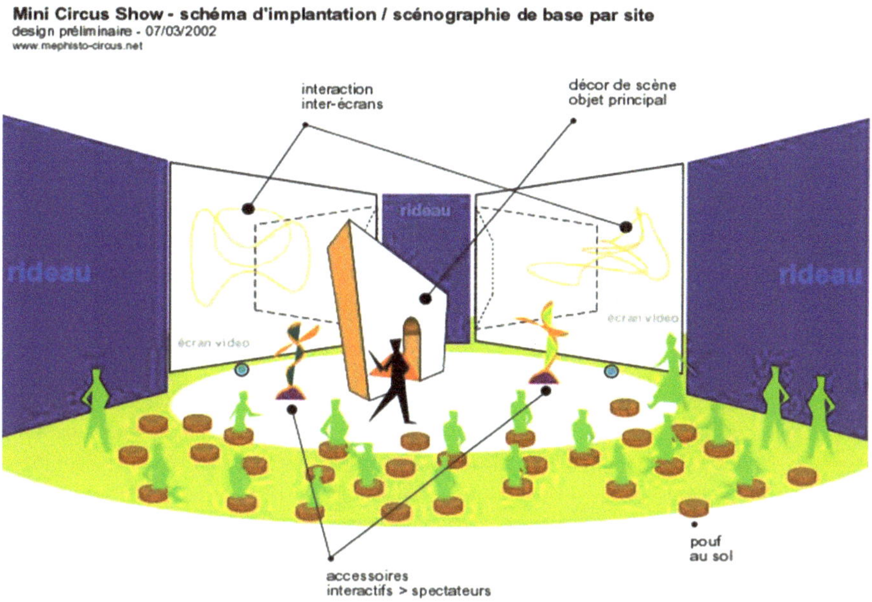

Fig. 5.13 Mephisto – Director file

or the participation of spectators, etc. Consequently, we need first to better qualify what these "tools" are. In fact, following Genette, Souchier and Jeanneret, I think that we need to get rid of the word "tool" and rather use the concept of "architext" to describe the pragmatics of these writing activities. Then, we need to look at the whole situation as yet another meta-system of writing that organized the confrontation of different versions of the show. Writing was not so much about using one particular software to achieve a goal but using a constellation of architexts that contrasted different visions of a show.

5.3.3.1 From Tools to "Architexts"

Design research has focused on design tools used at different stages of the design process: for instance sketching as shown by Schön and Wiggins,[45] Goldschmidt,[46] Kavakli and Gero,[47] or rapid prototyping.[48,49] The relevance of computer assisted design is also questioned for example during conceptual design.[50] Here, I want to

[45] Schön and Wiggins (1992).

[46] Goldschmidt (1994).

[47] Kavakli and Gero (2001).

[48] Sass et Oxman (2006).

[49] Cuff (1992) .

[50] Bilda et al. (2006).

consider the semiotic and pragmatic properties of the softwares and study them through the concept of "architext".

The word "architext" was first used by the literary theoretician Gerard Genette[51] (1979), and describes rhetorical patterns that underlie a group of texts. Each text is built and understood with explicit quotations, but more subtly because formal, stylistic, and rhetorical characteristics help the reader to recognize the genre of the text. Under the diversity of styles, the reader can perceive a relationship between texts that does not make them equivalent but similar. The focus here is on categories of texts rather than details between specific texts. Genette's concept takes into consideration the editorial dimension of these texts, the way each social group classifies them (prose versus poetry, for instance), gives them a status (commercial versus informational), legitimize them (legal or ludic).

Second, the depth and freedom of the interpretation depends on the culture of the reader. The architext may not be perceptible to a reader of elementary competence, like a kid who learns to read. A more confirmed reader recognizes the architext because she captures some of the stylistic characteristics that make it part of a genre. This competence is part of the "re-creating experience" described by Panofsky: "the recreating experience of a work of art depends on the natural sensibility of the spectator, on her visual training, but also her cultural background and experience[52]". Panofsky contrasts two relationships to works of art: anybody can appreciate the aesthetics of a work of art, but the recreating experience is based on an ability to compare it to others and to replace it at the time of its creation. The more skilled spectator will of course enjoy the experience but she will also judge its material characteristics, its qualification as an object of contemplation as well as some of the institutional dimensions of the work of art, and its place and role in the dynamics of art History.

The concept of architext has been used again with a slightly different meaning by Jeanneret and Souchier[53] as the tangible and visual pre-organization of a text as it appears in the window of laptops, with signs, tabs, and a model of text (typically with a professional feel and not a rough copy). The authors point out that the etymology of the word encompasses two ideas: first a beginning (the writer is not confronted with a white page) and second management (the process is spatially and temporally organized). The digital architext structures the practice of writing.

Our experience of writing and rewriting Mephisto with different "architexts" showed how these intellectual technologies provide resources to imagine a new text. At the same time, these architexts enclose the writer in a framework that can be detrimental to the ideation process because they incorporate models of what a proper text is (professional typewriting and publishing industry versus private sketching space, for example).

Of course, privileging one writing tool has never prevented authors from exploring different genres. In our case, what is at stake is not only the form of the docu-

51 Genette (1979).

52 Panofsky (1969).

53 Jeanneret and Souchier (1999).

ments but also the definition of what a show is, as well as the relationships between a show and texts. In this example, designing means considering what these tools mobilize in terms of models and how they can orient and disorient, but also how they need to be used together to actually organize yet another confrontation. The concept of architext also helps us consider that tools engage a representation of knowledge and defend a certain aesthetic. As we have seen, each architext offers a different model of writing, but also a different distribution of power as each of them emphasizes or downplays the role of actors, authors, props, technologies. In other words, they are part and parcel of an apparatus as it is discussed by Foucault. I will come back to the concept of apparatus at the end of this chapter.

5.3.3.2 Multiplying Architexts as a Writing Strategy

In the debate between those who design their own tools (as John Maeda or Antoine Schmitt presented earlier[54]) and artists or designers who use a tool of their choice and consider that the outcome matters more (Agnes de Cayeux working in Second Life for instance[55]), a third path advocates using several architexts with a critical distance, as was obviously Jaffrennou's position. The "constellation of texts" was a strategy to "de-naturalize"[56] (Barthes) the implicit model of text that is built within the software. The manipulation of different tools destabilized each result by a new one. What I therefore learned was that using different architexts was not necessarily a strategy to avoid the limitation of one of them, but rather to redefine what a theatrical text is. First, each tool produced different versions that not only built the show differently but also questioned the group's presuppositions of what a show "is". Each version was as valid as the other. It is worth noting that some theatre creators, like Jaffrennou but also Jean-François Peyret,[57] exhibit these documents not as sketches and drafts, but as creative spaces as well. They want to make a point that there is a variety of "works" and not a hierarchy or a succession of works leading to the triumphant "final" result: the show.

Second, revisiting a text with a different tool is a way of expanding how we think about contents and media (including the show) in relation with others. The emphasis here is on inter-textuality: the interdependence of texts creates a rich structure of evocations, contrasts, nuances, that echo each other and expand our perception of each. For example, the participants interpreted their first texts in Word differently after producing the Excel spread sheet. Each document influenced the interpretation of the others and the contrast of texts brought forward new unexpected interpretations and potential productions. It is a learning and expansive process.

---

[54] http://www.gratin.org/

[55] http://www.agnesdecayeux.fr/

[56] Barthes (2012).

[57] http://www.theatrefeuilleton2.net/, http://fr.wikipedia.org/wiki/Jean-Francois_Peyret

5.3.4 Designing a Field of Tensions to Fight the Apparatuses

In both examples, the creators did not only design something new, they redesigned the situation into a confrontation so as to expand the conception of their "product". In the first case —VUE, the constitution of a double corpus supported the defixation from the techno-semiotic characteristics of a communication platform and helped redefine e-learning solutions. In our second case —Mephisto, the diversity of tools provided a diversity of texts and viewpoints that allowed the theatrical production to be redefined. These operations were part of a design situation of confrontation that structured the dialog with models of texts and situations.[58] In other words, tensions appeared between versions, visual grammars, contents, and through them, different systems of values, different perceptions of interactions, different worldviews. Understanding these situations of confrontation can therefore help us understand the human, social and technical interplay in design but in this section I want to emphasize their aesthetics goals and the role of composition.

The concept of "dispositif" or "apparatus" as it was developed by the Italian philosopher Giorgo Agamben[59] after Foucault, can help us focus on the design situation. The word "dispositif" was used by Foucault to describe sets of human, tangible, even architectural constraints that organize the way people live and produce. An apparatus is both a way to organize knowledge in different fields (psychology, psychiatry, medicine, etc.) and to organize power (a prison, a hospital, a church, etc.). In a discussion published as "The Confession of the Flesh" in 1977, he describes the dispositif as: "a thoroughly heterogeneous ensemble consisting of discourses, institutions, architectural forms, regulatory decisions, laws, administrative measures, scientific statements, philosophical, moral and philanthropic propositions–in short, the said as much as the unsaid. Such are the elements of the apparatus. The apparatus itself is the system of relations that can be established between these elements."[60] His definition focuses on the strategic elements of such configurations: "which means that we are speaking about a certain manipulation of relations of forces, of a rational and concrete intervention in the relations of forces". Agamben further employed the word to mean:

> literally anything that has in some way the capacity to capture, orient, determine, intercept, model, control, or secure the gestures, behaviors, opinions, or discourses of living beings. Not only, therefore, prisons, madhouses, the panopticon, schools, confession, factories, disciplines, judicial measures, and so forth (whose connection with power is in a certain sense evident), but also the pen, writing, literature, philosophy, agriculture, cigarettes, navigation, computers, cellular telephones and--why not--language itself, which is perhaps the most ancient of apparatuses--one in which thousands and thousands of years ago a primate inadvertently let himself be captured, probably without realizing the consequences that he was about to face.[61]

[58] de Grazia and Furlough (1996).

[59] Agamben (2009).

[60] Foucault (1980).

[61] Agamben (2009).

An established genre (like the e-learning websites in VUE) or a writing tool are therefore apparatuses that structure the way we think about learning or writing. This has important consequences on how we consider certain matter of fact operations such as making "a state of the art". When researchers and designers build a state of the art of existing corpuses they augment their knowledge of the field. However, they also reinforce the evidence of the apparatuses and fix the representations of the common, usually admitted aesthetics traits of this field. Similarly, architexts not only support tasks. As apparatuses, they also create blind spots. The design space consists therefore in clashing these different apparatuses to challenge the way we relate to tools and texts and how we define ourselves in relation to them. This second aspect is fundamental as it focuses on the designer's practice of tools: should it be one of mastery? Or one of play?

Agamben's definition is also helpful in that it recognizes a double relation to the apparatus that was actually very much felt by the participants in the writing process of Mephisto. Each tool/apparatus helped the participant to write and, therefore, expand their subjectivity. For each of us, it was not only about inventing something it was also a personal journey of self-discovery. Agamben points out that indeed this interplay between the living actor and the apparatus defines the subject. In other words, the interaction between living and nonliving actors (to use the Actor Network Theory vocabulary) builds the subjectivity. As designers, it was important to let ourselves be influenced by the tools that we used, rather than to try and control them. We needed the tools to surprise us in shaping us in particular ways. In his creative writing workshop, Jaffrennou let the production be led by the inherent design of each architext. He let the participants be played by the architexts. But at the same time, he created a system of confrontations to counter the limiting effects of each architext/apparatus. Had we used only one tool in confidence that it could support our whole creative process, we would have been closed into one model of thinking about the theater. To organize a situation where there is not one but several architexts was to painfully recognize that subjectivity has to find a way in between all of these tools. Whether through a contrasting semiotic analysis or through the use of several architexts, the designers not only manage a project but also let the situation, artifacts, tools, shape the production by exposing the multiple facets of a new media, service, or artform.

The concept of "dispositif" also foregrounds the questions of format, shape, and, more generally speaking, of aesthetic composition. In Agamben's definition of the "apparatus",[62] which is yet another way to translate the French "dispositif" and the Italian "dispositivo", this interplay of structure and people also creates specific aesthetics. In Latin, the word "dispositio" means the composition. While only one tool or one corpus of references would have limited the composition, bringing in more tools and sources opened up the elements to compose with. Composition here is not only the strategic guiding of sources and tools it is also how we are led by these elements, their here and now "material agency" as Knappett and Malafouris[63] have

[62] Agamben (2009).

[63] Knappett and Malafouris (2008).

elaborated. From this perspective, the "idea" – that is the starting point in most design models – is obtained at the end, after the confrontation of media and the use of multiple tools and not at the beginning. This is an important aspect of a model of design as a plane of composition. Through confrontation, designers engage in a dialog with previous artifacts as sources for the composition. The contrasting semiotic analysis as well as the constellation of tools, but also moodboards,[64] or materiautheques[65] are a deliberate organization of the composition space, a net to catch the elements that might lead to a new design. They build the matrix that will make it possible to create something new. To think about the design process as systematically starting with the ideation phase then could seriously be misconstrued.

A serious criticism of the starting "idea" is presented in Henry James' short story: "The Figure in the Carpet" (1896). In this short story, Henry James comments on artistic work with an underlying critique of what people expect it to be: that is an idea well performed. The protagonists of the story therefore look for an idea – hidden but at the origin of the books – an idea so powerful that it could change their lives. A young critic (the narrator of the story) meets the writer so as to ask him about this idea. The writer does not want to answer but repeats that the answer is in the several volumes that he has already written: "It stretches, this little trick of mine, from book to book, and everything else, comparatively, plays over the surface of it. The order, the form, the texture of my books will perhaps some day constitute for the initiated a complete representation of it". Before Duchamp, at approximately the same time as Paul Klee, Henry James loses his readers in a quest for meaning and lets us know that the work of art is the work of art, without precedent, but each time as it is recreated by the viewer. As we read the short story, wondering about its meaning, we forget to look at what is there: the work of words. The use of the carpet metaphor is not innocent. James has an idea of art as weaving things together. The idea emerges from the weaving, not the weaving from the idea.

Similarly, in our two cases, the starting point is vague and the ideas and the productions emerge together gradually through the confrontation of materials. The "image of the text" is, therefore, not only an idea put into a shape but a gradual building of an aesthetic that is material and ideological both and at the same time. Such viewpoint is sustained by the anthropologist of material culture, Tim Ingold, who, elaborating on Klee, also speaks of the "The Textility of Making"[66] that I further discuss in the next section.

5.4 Conclusion: "Two to Start"

Many creative activities and situations described by artists, designers, or even engineering researchers, undermine the managerial model of design where linear time rules design organization as a sequence of events, and where each activity feeds the

[64] Gentes et al. (2015).

[65] http://www.citedudesign.com/fr/materiautheque/

[66] Ingold (2010).

next one. We need to switch from a chronological model of design to a spatial model of design where the focus is on the elements in presence, in the situation, where all the actors living and nonliving are being composed to beget a new unknown.

5.4.1 *Going from a Metaphysics of Design to a Pragmatic of Design*

The philosopher of design, Pierre-Damien Huyghe, in a small and interesting book called: “Commencer à Deux” – “Two to start”[67] – analyzes the fact that since Aristotle we consider design – or architecture – as a linear process starting with an idea that eventually leads to its implementation. Looking at the word architecture and at how it is considered in the Occident, he points out that in this one word there is:

- the root: “archi” the starting point, the idea that is also the archive,
- and the final touch: the roof that the carpenter puts on a building. In other words, the idea leads the whole construction process.

In this model, the primary idea is all that matters because it is all that is needed for the project to be fully archived. Huyghe qualifies this as a metaphysics of design, that is a theory based on a representation of ideas as holding all of the creative power. Indeed, in Nicomachean Ethics,[68] Aristotle describes the three stages that deploy this process from abstract to tangible. The first is the conception that is a way to consider what “could be”, what we would call now ideation. The architect is the perfect embodiment of this stage. Around 15 BC, Vitruve, in De Architectura,[69] considers that the essence of architecture is indeed to contemplate the possibility of something.[70] The second stage is about building, that is to say to pass from the generality of the conception to the particular case (what today such researchers in design as Willemien Visser[71] after Tulving and Thomson[72] qualify as “episodic knowledge”) that is to say the capacity to re-use some experience of the same kind. Finally, the third stage is about the “know how”, the craft of skilled workers. What we just described is the very model that the industry has implemented in its hierarchy, organization, and processes. While the model has its merits, Huyghe, not without a sense of humor, suggests that we should consider design, or conception, as the meeting of two sources that would otherwise produce nothing on their own. The birth of an unknown might not be the result of this metaphysics of conception but a far more tangible meeting of parts, here and now, that beget the invention. To sup-

[67] Huyghe (2009).
[68] Aristotle (1999).
[69] Vitruve (1995).
[70] Ibid.
[71] Visser (2006).
[72] Tulving et Thomson (1973).

port his criticism of the Aristotelian model of design, Huyghe borrows his vision from one of the founders of the Bauhaus, Paul Klee, who in "On Modern Art",[73] suggested that we should move our vision of design from "the model" to "the womb of nature, at the source of creation". In Klee's words the "model" depicts industrial design, that is a chronological process starting with an abstract idea that gradually takes shape and is materialized through sketches and prototypes. But he suggests that design can also be analyzed as a platform supporting a palette of materials, colors, shapes, and a variety of processes. In his view, the emergence of ideas and shapes happens "at the same time" through the frictions between the elements that the designer brings together. Design in this instance is more about the organization of confrontations of material elements. Design, therefore, is not only project based. It depends on a matrix, where hybridization can take place, through rejection, overlap, aggregation, etc. The two views are probably complementary. A design project does start with a briefing that different stakeholders discuss and through different stages brings about the finished product. But these discussions and stages are also supported and contrasted, in many instances, by getting bits and pieces of material, images, texts, building elements, drawings, schemas and the presence of multiple tools. One way to look at this practice of discourses and drawings is to say that it helps to embody one precise idea, another is to say that it explores shapes and lets new ideas emerge.

The anthropologist Tim Ingold, also elaborating on Klee, further develops this discussion by pointing out that a model of creation as a "matrix" pays attention to different aspects of the creation: in particular, the forces and materials that are shaping the ideas.

> Contemporary discussions of art and technology continue to work on the assumption that making entails the imposition of form upon the material world, by an agent with a design in mind. Against this hylomorphic model of creation, I argue that the forms of things arise within fields of force and flows of material. It is by intervening in these force-fields and following the lines of flow that practitioners make things. In this view, making is a practice of weaving, in which practitioners bind their own pathways or lines of becoming into the texture of material flows comprising the lifeworld. Rather than reading creativity 'backwards', from a finished object to an initial intention in the mind of an agent, this entails reading it forwards, in an ongoing generative movement that is at once itinerant, improvisatory and rhythmic.[74]

Huyghe and Ingold therefore not only offers new concepts (also taken up by Deleuze and Guattari in A thousand Plateaux[75]) but suggest a program for designers and human scientists that we want to focus on in the next section.

[73] Klee (1966).

[74] Ingold (2010)

[75] Deleuze and Guattari (1987).

5.4.2 The Art of Composing

Multiplying ideas and confronting alternatives are a way to trigger crises and "surprises" that create new questions and framings. While it can be left to chance, most designers will organize a situation so that it does not give one "straight answer" but provides many options that challenge pre-existing conceptions. That is exactly the point Schön makes when he describes a design situation and comes up with a model of design as a conversation with tangible artifacts, situations, representations, even the designer's body. While Schön speaks of "transaction",[76] I suggest that composition might help us better understand the way that the process of confronting elements actually brings a new artifact.

There is no art without composition. Composition organizes how parts are being assembled, the relations between the parts, but also the relations between lines, between light and dark areas, between colors, textures, sounds, etc. While artists have always composed, Rosenberg[77] explains that the concept itself is quite recent. The first autonomous treatise on composition in painting dates back to 1784: *Saggio sulla composizione della pittura*, by Baldassarre Orsini. This book is a full-fledged treatise that put together analyses of works of art and recommendations on how to structure clair/obscure, foreground / background figures, perspective, etc. Orsini was elaborating on Alberti's *De pictura* (1435), who described the process of painting as:

> We divide painting into three parts, and this division we learn from Nature herself. As painting aims to represent things seen, let us note how in fact things are seen. In the first place, when we look at a thing, we see it as an object which occupies a space. The painter will draw around this space, and he will call this process of setting down the outline, appropriately, circumscription. Then, as we look, we discern how the several surfaces of the object seen are fitted together; the artist, when drawing these combinations of surfaces in their correct relationship, will properly call this composition. Finally, in looking we observe more clearly the colours of surfaces; the representation in painting of this aspect, since it receives all its variations from light, will aptly here be termed the reception of light. Therefore, circumscription, composition and reception of light make up painting;[78]

In this model of design, the work of art does not come from an idea that is slowly implemented, it comes from these three operations: delineation/circumscription, composition, reception of light. The first step is "circumscription". Circumscribing is a two-step process. Alberti came with the metaphor of the window because the painting sets a delimited space that structure the representation and the way people will look at the painting. The work of art therefore depends on the definition of a territory: not only the canvas and frame but also all the elements that are going to be used for the composition. The second stage is composition. Composition has to do with the *de facto* surprising arrangement of elements. The third stage is the reception of light. It refers to the appearance of the elements and therefore stands in between the choices of the media and the conditions of reception. The three operations are totally interdependent and of course have been interpreted in multiple ways

---

[76] Schön (1992).

[77] Rosenberg (2008).

[78] Alberti (2013).

up since Alberti. However, even in the most contemporary forms of performance[79] and diffusion (the post office for "art letters"[80]) the operations still define how the creative space is built and how it balances the elements of the composition.

5.4.3 *From Inductive/ Deductive Methods to Projective Abductive Methods in Design*[81]

In this section, I want to discuss some of the epistemological differences between a model of design as a project and a model of design as a composition. In my opinion, these models endorse different ways of building knowledge along with making new artifacts.

In addition to its practical advantages in the industry, the theoretical strength of the model of design as a project is that it can actually benefit from an inductive/ deductive methodological structure. The design project appears as a hypothesis derived from observations and the analysis of existing situations to deduce some unforeseen practice or aspiration that can then be changed, augmented, fulfilled. The hypothesis is implemented in a demonstrator that is tested and that produces new knowledge. In other words, the hypothesis is confirmed by the designed object. On the contrary, the model of design as a composition of tensions puts in the forefront abductive methods. I will try to show the rationale behind such a view of design, a view that seems more and more to be shared by researchers studying design activities. For example, it appears in Liam Bannon's[82] summary of the evolution of Human computer interaction in the industry. He first describes deductive phases, where the problem is known, and the process consists of verifying the design hypothesis through tests. He also describes inductive phases, where the designers gather information about users to understand their behaviors and come up with design question. Finally, he emphasizes an abductive phase where the designers' contribution consists of looking for clues and making unusual connections.

While induction and deduction have been discussed in epistemology of sciences, abduction has received less attention, but is now the subject of a renewed interest in design to explain "lateral thinking", free associations, hypotheses, and more generally projection in the design work.[83,84,85] These activities are part of the meaning-making process at play in design as studied in design semantics[86] or design

[79] Feuillie (2002).

[80] Saper (2001).

[81] This part of the chapter relies on workshops on Peirce's semiotics organized with Camille Jutant, Mathias Béjean, and Cedric Mivielle.

[82] Already quoted in Chap. 3 to address the question of the user.

[83] Roozenburg (1993).

[84] Sowa and Majumdar (2003).

[85] Schurz (2008).

[86] Krippendorff (1989).

semiotics.[87] Coming from design and innovation research, Buxton,[88] Kelley,[89] and more recently Dow et al.,[90] have also pointed out how parallel design supports learning and innovation. Amongst these different analyses, I find Chow and Jonas' demonstration which is explicitly based on Peirce's semiotics and theory of logic particularly useful in particular because they focus on "creative abduction",[91] that is abduction turned towards the possibility of something rather than the discovery of some hidden connections. Design/practice includes a sequence of activities: observing, reflecting, deciding and acting. Jonas points out that these activities involve three different types of knowing: analysis, projection, and synthesis. What is questioned is the very sequence of this macro process. Nelson and Stolterman[92] consider that though analysis enriches the design solution it does not "cause" design. More to the point, the idea that an analysis of the situation precedes the design itself is related to an idea of design as problem solving. As we have seen, if design is problem solving, then identifying all traits of the situation is necessary to the design process. But if design is seen as an expansion of the real[93] then what matters more is a domain of knowledge (for example teaching) and a series of concepts that challenge the situation as it is traditionally understood. Chow and Jonas contend that "existing artifacts are knowledge sources for projection of the new".[94] They qualify as "transfer" the fact that "we can take knowledge from one artifact and put it in another domain or context to create something new".[95] In the e-learning case—VUE, for instance, this transfer occurred at several levels: it worked on the form of the service (as it recognized similarities), on the context of the service (as it took from one context to place in another) and on the underlying design principle (to share someone's point of view). From their perspective, transfer is related to Peirce's theory of sign and meaning making, and more specifically his theory of abduction.

It is actually difficult to find a definitive version of what Peirce meant with abduction in his writings as he produced several examples and explanations. The first definition of Pierce's abduction is that it recognizes a hidden relation between two elements.[96]

> All that makes knowledge applicable comes to us via abduction. Looking out of my window this lovely spring morning I see an azalea in full bloom. No, no! I do not see that; though that is the only way I can describe what I see. That is a proposition, a sentence, a fact; but what I perceive is not proposition, sentence, fact, but only an **image**, which I make

[87] Chow and Jonas (2010).

[88] Buxton (2007).

[89] Kelley (2002).

[90] Dow et al. (2010).

[91] Eco and Sebeok (1988).

[92] Nelson and Stolterman (2012).

[93] Hatchuel and Weil (2002).

[94] Chow and Jonas (2010).

[95] Chow and Jonas (2010).

[96] I would like to thank Warren Sack for his judicious remarks and discussion on Peirce.

> intelligible in part by means of a statement of fact. This statement is abstract; but what I see is concrete. I perform an abduction when I so much as express in a sentence anything I see. The truth is that the whole fabric of our knowledge is one matted felt of pure hypothesis confirmed and refined by induction. Not the smallest advance can be made in knowledge beyond the stage of vacant staring, without making an abduction at every step.[97]

Here, abduction means uncovering relations that were already there. Hence the comparison between abduction and a detective enquiry that puts together the different clues to solve a mystery.[98] As emphasized by Warren Sack in our discussions, Peirce says in its most basic form that abduction is guessing. Peirce wrote, "Abduction is no more nor less than guessing,...".[99] This first definition of abduction leans towards a static and not expanding world of signs. However, Schurtz analyzes that there is a major difference "between selective abductions, which choose an optimal candidate from given multitude of possible explanations, and creative abductions, which introduce new theoretical models or concepts".[100] I would like to elaborate on abduction as a dynamic production of new meaning based on the quality of things (Firstness), something that I want to call "projective abduction" to mark the creative nature of the cognitive operation. In the next section, I am therefore less interested in the signs proper and more in the operations that lead to the signs.

5.4.4 *"Projective Abduction"*

I will not sum up here the whole of Peirce's theory.[101] However, I want to follow a few threads to understand how the composition is based on "projective abductions". In the case of deduction, a law is imposed on things. This law is a social phenomenon as it is fully stated in symbolic terms in the linguistic form of a hypothesis. Deduction is therefore a process based in Thirdness, that is a plane of meaning that relies on socially shared knowledge. Thirdness in Peirce's philosophy, is the category of language and representation which makes social communication possible. In the case of induction, the elements are reduced to the symptoms of a law. Induction depends on Secondness as it observes events, objects, here and now. Induction means that the observer looks at things without yet coming with a socially sharable theory or hypothesis. It is essentially the plane of practical experience and the plane of elements in action-reaction, of witnessed causes and consequences. In the case of abduction, there is the idea that elements could have a meaning if taken together. What I find interesting at that point, is that the abduction process is intuitive

---

[97] 'The Proper Treatment of Hypotheses: A Preliminary Chapter, toward an Examination of Hume's Argument against Miracles, in its Logic and in its History' (MS 692), HP 2:899–900, 1901).

[98] See for example, Harrowitz (1984).

[99] Prolegomena for an Apology to Pragmatism, (MS 293), NEM 4:319–320, c. 1906.

[100] Schurz, « Patterns of abduction ». p. 201.

[101] For a first introduction to Pierce's semiotics, see http://www.signosemio.com/peirce/semiotics.asp

and based on feelings of a possible connection. It is a perception of quality that starts an interpretive process. Abduction is therefore based in Firstness, that is the plane of the possibility of something, the experience of a latent potentiality. Firstness is detached from the actual practical experience, or the social experience. It is a subjective experience that is not yet embodied in a full recognition of elements in tension, nor in a shared social rule or habit, or law.

I find that Shank's reformulation of Peirce's categories of sign[102] is interesting in that it translates Pierce's terminology in expressions that give a vision of design practice. Peirce also used some of these words as alternatives to his final terminology. Here is Shank's model with Peirce's final terminology in brackets.

- Open (rhematic) Iconic Tone (qualisigne)—**hunch**
- Open (rhematic) Iconic Token (sinsigne)—**omen**
- Open (rhematic) Iconic Type (légisigne)—**metaphor**
- Open (rhematic) Indexical Token (sinsigne)—**clue**
- Open (rhematic) Indexical Type (légisigne)—**pattern**
- Open (rhematic) Symbolic Type (légisigne)—**explanation**
- Singular (dicent) Indexical Token (sinsigne)—**fact**
- Singular (dicent) Indexical Type (légisigne)—**hypothesis**
- Singular (dicent) Symbolic Type (légisigne)—**theory**
- General (argumental) Symbolic Type (légisigne)—**demonstration**

The "open" or "rhematic" signs are those that play on a latent potentiality (Firstness). Here I suggest that this potentiality is not only something that "might" exist in relation to a present experience but something that "could" potentially exist. "Open" or "rhematic" signs explore possible futures. In other words, the potentiality affects what could happen in another place or another time.

What matters is the fact that the first three signs – "hunch, omen, metaphor" – as they are reformulated by Shanks – are iconic. An iconic sign is one that has a relationship of similarity from a certain angle to its object. The "hunch", to follow Shank's terminology, is the first inkling that things could be connected. At the beginning of the enquiry, the observer notices things that could be related to a possibility by similarity. If we go back to our use case VUE and the contrasting semiotic analysis, this is precisely this "hunch" which was at play. There was the hunch that the point of view in cinema would somehow be similar to the aesthetics in e-learning platforms. In the use of architexts for the Mephisto show, the different written versions were also deemed to be somehow similar with the potential future show.

Shank then goes on to describe the omen as "a sign of the possibility, based on current resemblances, **of a future event**". Here the relation to design is quite obvious. Abductive reasoning is about things that are not there on a certified basis but that could happen. If we consider design as a reasoning that precisely builds on the potential to actually produce new artifacts, then the omen as a projection in the future is a necessary abductive function of design reasoning. In our two use cases,

[102] Shank (2001).

the potential for new forms of representation and performance was at the basis of the whole design activity. In Jaffrennou's case the writing process that confronted different tools was announcing a future show that was not precisely defined. In the e-learning case, the comparison between different media was not made for comparison's sake, but to see through the analysis the chance of another form of representation.

Shank also describes the "metaphor". His works leads him to assess different degrees of resemblances that range from being identical or equivalent to looking alike: "When we reason to a metaphor we are deliberately manipulating this tension between equivalence and resemblance. That is, a metaphor is stronger than a resemblance claim but weaker than an equivalence claim [...] a metaphor is a rule or law based on nothing other than possibility".[103] The design practices that we observed were indeed considering options by creating a surprising confrontation of semiotic systems based on a metaphoric process. In the e-learning use case, the team created the conditions for abductive thinking first by analyzing the semiotic characteristics of existing systems then by comparing them to other semiotic systems from different media. In the theater case, the use of different tools while working on the same "theme" produced the same kind of confrontation. This metaphorical process presupposes similarities and differences. I will further develop how the presupposition is made in the next section.

Considering "projective abduction" as the way we elaborate on future possibilities helps us recognize the meaning of the field of tensions and the way the composition works. Designers do not make a collection of data and materials for the collection's sake. They need to force the chance encounters of different elements, to "perceive an image", to build a new meaningful artifact bearing enough resemblance to other activities and artifacts while detaching itself significantly from the rule (Thirdness). Firstness as a feeling of possibility is therefore specifically provoked by practices that are built from the confrontation of elements proper to the Secondness of Peirce's theory. The general potential "sensemaking of our world" depends on the capacity of designers to make connections where none primarily exist or none are even imagined. But contrary to our everyday relation to the world, or a scientific approach to surprising events, abduction in design is carried a step further because it is provoked by a specific organization (semiotic comparisons, moodboards, various writing tools, etc.) that lets the imagination project new possibilities, creates its own surprises, and retroactively finds meaning for them. In other words, it tries new combinations and forces the mind to exercise its creative power of interpretation to address the new form. Organizing a confrontational dispositive as a matrix for future design therefore consists in preparing a situation, or building intellectual tools that support projective abductive thinking. Contrary to the CK theory elaborated by Weil, Hatchuel, and Lemasson, and later described, that starts with a concept, the theory of the matrix and projective abduction starts with materials and their aesthetics. In the realm of language as a poetic material, the metaphoric process analyzed by Eco and Paci describes how projec-

[103] Shank Ibid.

tive abduction is both "crazy" and finally coherent. Elaborating on their analysis, I want to close this chapter on paradox and coherence.

5.4.5 "The Earth Is Blue Like an Orange". The Claim to Paradox and Coherence

"The earth is blue like an orange" writes Paul Eluard[104] who goes on saying: "Never an error, words do not lie". This poem is for me the archetype of what I understand about the double claim of an original composition. Less talented, but working in the same way, "the e-learning situation is the loneliness of the long distance runner", or "the show is an excel sheet", are all oxymorons (breakthrough) that end up in metaphors (coherence), in other words impossible associations that finally make sense on a certain level because they redefine the way we think about the earth, the orange, learning, or the theater: "Never an error, words do not lie", that is the claim of any original associations that want to say something true and meaningful from a certain vantage.

The semiotician, Umberto Eco, notes that the metaphor simultaneously exploits similarity and difference not from an ontological point of view (that is not because the elements of the metaphor have some real common features), but from a semiotic point of view. In their article on "The scandal of metaphor[105]", Eco and Paci retrace the different perspectives about metaphors since Aristotle. They demonstrate that the metaphor is a semiotic process where two elements work paradoxically because they must have enough similarity to place them in the same paradigm, but enough difference for the comparison to have the necessary element of contrast.[106] The metaphor therefore changes both initial elements as in a "condensation" process (Eco and Paci use Freud's terminology about the interpretation of dreams[107]) where the original elements are transformed in the work of the dream. Umberto Eco points to the logical process at play in particular as a metaphor is grounded in what he calls the encyclopedic meaning-making process. He explains that in the semiotic process we undertake an impassioned "hermeneutic circle":

> One assumes a code, which is verified against the simile, whose metaphorical transformations are appraised in advance; or one starts from the simile in order to infer a code that makes it acceptable; [....] Analyzing further this process of trial and error, we would realize that we are dealing with multiple inferential movements: hypothesis (or abduction), induction, and deduction.[108]

[104] Eluard (1966).

[105] Eco and Paci (1983).

[106] Warren Sack remarked that Amazon.com's recommendation system works on the same principle when it is matching one buyer's profile to its database of profiles in order to suggest other similar books to buy.

[107] Freud (1997).

[108] Eco and Paci (1983).

The metaphorical process in Eco and Paci's definition "posits" (in a philosophical sense, but also in a physical sense, as in "putting before the eyes") a proportion that is unexpected. It is like an oxymoron: a figure of speech in which apparently contradictory terms appear in conjunction. The composition starts a process of interpretation that actually builds new meaning to make sense of it. In our examples, the design of the artifacts somehow demonstrates that the oxymoron is not only aesthetically interesting but cognitively valid. The design validity is discovered after the fact through a process that Eco and Paci call the "Porphyry's tree", named after its author, neoplatonist philosopher and logician Porphyry.[109] The Porphyrian tree is the representation of the logical path of a metaphor that can be claimed after the new object is produced. But rather than considering it from the standpoint of an already existing and discovered relationship, I think that we need to consider the process as the invention of a future potential relationship.

At this point, I would like to go back to the two use cases of this chapter and show how the design solutions, that seemed paradoxical, actually could be presented as converging at a certain level. I am interested in getting to understand how a seemingly "crazy concept"[110] can claim a rationale by reorganizing two knowledge bases as the CK theory demonstrates.

Paradoxical metaphors (or oxymorons) were at play in both use cases as a way to merge seemingly divergent propositions. In the theater case, the artist implicitly told the group that a show is like an html tree, or like a spreadsheet. Excel is a spreadsheet to execute operations while a play is actions, people, props. There is obviously no common point. But the metaphor still works because it merges at a certain level as shown on the following diagram. It works because on the one hand, it reduces excel to its aesthetic qualities: it is a grid that organizes data through a matrix. It also works because it narrows down the scope of the theater by omitting the narrative structure of a theater play and by limiting it to a set of actions. At the same time, it enlarges the purely accounting vision of excel as a spreadsheet by giving it a creative capacity and it augments the perception of the theater by giving the same attention to things as to people and actions. Finally, the oxymoron induces a reflection on the mechanisms of the theater by focusing on its malleability and playful nature. The coherence is not pre-existing but is built through a reorganization of what Hatchuel et alii call the knowledge base (Fig. 5.14).

The same metaphoric process was at play in the abductive reasoning of the e-learning case. The intuition of the design group was to focus on how to provide a rich visual experience that could counterbalance the effects of the students' loneliness. They used the metaphor of the "control room" as defining the learning class. But between cinema – or live TV feed – and a learning environment, there is very little in common. Again, for the metaphoric process to work, it had to narrow down and converge the new definition of cinema and of e-learning platforms. On the one hand, cinema was considered in relation to its contribution to image syntax: close-up, wide angle, shot/reverse shot, etc. The design team also questioned how such syntax could be done "live". They subsequently followed that lead up to the TV control room. On the other

[109] http://en.wikipedia.org/wiki/Porphyrian_tree

[110] Hatchuel et al. (2014).

hand, rather than listing all the tools available in e-learning platforms, they focused essentially on people's presence and forms of representations. The main image of e-learners is the front shot of their face, provided of course by their webcam. This image seemed like a reduction of the variety of angles and views that an actual class provides. The metaphor came as a visual solution to a participatory question (Fig. 5.15).

In both cases, the metaphorical process is substantiated by a claim of coherence that is apparent in the logical Porphyrian tree. The coherence is "guaranteed" by the change of meaning of both primary concepts that finally borrow from each and therefore expand beyond their original meaning. The spreadsheet is no longer considered only as a combinatorial tool that "treats" only numbers but also as a mechanism that combines actual things, people, situations, and therefore reveals their malleability. The theater is no longer considered as a story put on stage, but as a dynamic, compositional space, where everything contributes to the experience of the show. In the e-learning case, people are no longer seen as live individuals but as characters on a stage, who can suddenly direct theirs' and others' image. The Porphyrian trees are a way to analyze a claim to a rationale and to observe the reductions and expansions of meaning.

5.4.6 *Open Conclusion: Design as an Apparatus of Tensions*

As I started this chapter on design as composition of tensions, I evoked the students' experience of discomfort. This discomfort is not only a psychological consequence of innovation. It is the consequence of a semiotic process bringing into coherence seemingly different events, knowledge bases and patterns. The psychological key to a scientific behavior is the feeling of surprise that is related to the challenge of preconceptions. Peirce talks about "genuine doubt" and notes that this state is uncomfortable. He also notes that we generally try to "fix" it as soon as possible. The two

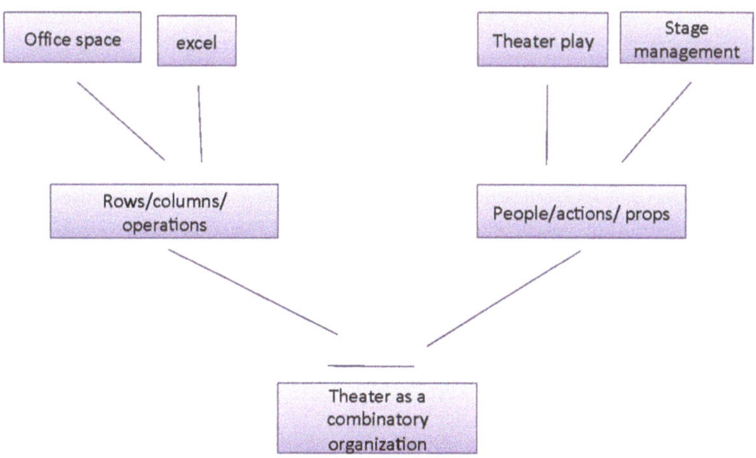

Fig. 5.14 Porphyrys' tree of the theater use case: Mephisto

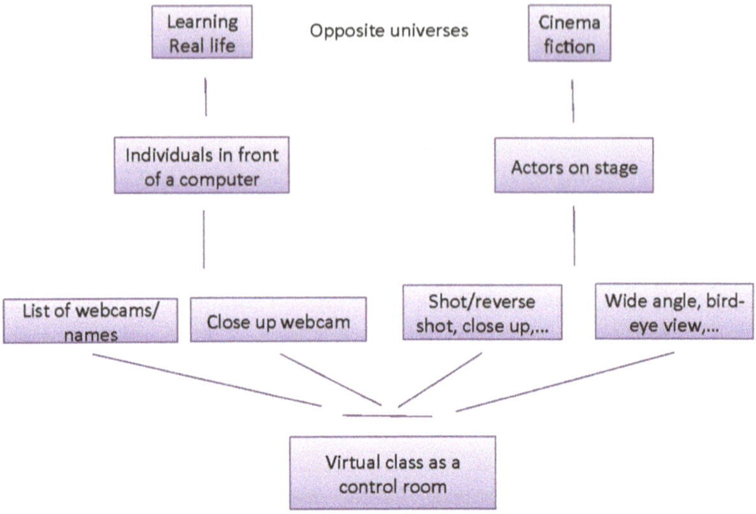

Fig. 5.15 Porphyrys' tree of the e-learning platform VUE

examples in this chapter describe two dispositifs of confrontation – two "matrices" to evoke Paul Klee's words – that triggered and organized such discomfort but also helped the design process. First, they were a way to generate divergent conceptual and aesthetic options, which is a crucial challenge in design work. The semiotic analysis of other media dealing with the same issue and the use of a constellation of writing tools were a way to reduce convergent thinking and break free from one single mind frame. The analysis of these situations led us to consider them as dispositives that structure not only knowledge but also power and aesthetics.[111] In these dispositives, the confrontations as they are embedded in the software or contents themselves, are the sign of an abductive process. From a design perspective, it seems important that abduction should also be seen as an aesthetic process. Because abduction is based on the open iconic semiosis, the design process is not only about ideas leading to forms but also forms leading to ideas. This challenges a general view of design organizations that focuses on a strict chronology of ideas, sketches, implementation, production, and tests. Organizing a field of tensions so that ideas emerge from materials can be just as important.[112]

What the notion of dispositive brings into the equation is the material conditions of such a form of reasoning: abduction is incarnated. It has to be explicitly embedded in situations that force design practice into abductive mode. In Shank or in Eco's analyses, the situation is mostly a given: the analysis in the detective story or the structure of a successful poem. But the situation is not a given in design: the use of multiple architexts or of contrasting semiotic analyses are only two examples of such organizations that need to be deployed when some design work has to be done.

[111] Catellin (2004).

[112] Dow (2010).

In addition, the situation is also made of other living actants, who more or less actively multiply the possible projections. Who are they? How do they contribute to the design space? What are the powers at play in a design situation? These last questions will be considered in more detail in Chap. 6 as we look at who participates in the design process and at how the "object" is also a "thing" that is debated.

References

Agamben, G. (2009). *What is an apparatus? and other essays* (trad: Par David Kishik et Stefan Pedatella). Stanford: Stanford University Press.

Akrich, M. (1990). De la sociologie des techniques à une sociologie des usages. *Techniques & Culture, 16*, 83–110.

Alberti, L. B. (2013). *On painting*. Cambridge: Cambridge University Press.

Aristotle. (1999). *Nicomachean ethics* (2nd ed.). Indianapolis: Hackett Publishing Company, Inc.

Armogathe, J., et al. (2001, Folio). *La Querelle des Anciens et des Modernes : 17e-18e siècles*, Reprint.

Artaud, A. (1994). *The theater and its double* (trad: Par Mary Caroline Richards). New York: Grove Press.

Aumont, J., and Marie, M. (2004). *L'analyse des films* (2e éd.). Paris: Armand Colin.

Aumont, J., et al. (1992). *Aesthetics of film*. Austin: University of Texas Press.

Barthes, R. (2012). *Mythologies. The complete edition* (trans: Howard, R., & Lavers, A.). Hill and Wang.

Bilda, Z., Gero, J. S., & Purcell, T. (2006). To sketch or not to sketch? That is the question. *Design Studies, 27*(5), 587–613.

Blandin, B. (2004). La relation pédagogique à distance, que nous apprend Goffman? *Distance et savoirs, 2*(2/3), 357–381.

Buxton, B. (2007). *Sketching user experiences: Getting the design right and the right design*. Amsterdam: Morgan Kaufmann.

Catellin, S. (2004). L'abduction: une pratique de la découverte scientifique et littéraire. *HERMÈS, 39*, 179–185.

Chow, R., & Jonas, W. (2010). Case transfer: A design approach by artifacts and projection. *Design issue, 6*(4), 9–19.

Cuff, D. (1992). *Architecture: The story of practice*. Cambridge: MIT Press.

de Grazia, V., & Furlough, E. (Eds.). (1996). *The Sex of Things: Gender and Consumption in Historical Perspective*. Berkeley and Los Angeles: University of California Press.

de Meredieu, F. (2005). *Digital and Video Art*. Edinburgh: Chambers.

Deleuze, G., & Guattari, F. (1987). *A thousand plateaus: Capitalism and schizophrenia* (trad: Par Brian Massumi 1 ed.). Minneapolis: University of Minnesota Press.

Doane, M. A. (2003). The close-up: Scale and detail in the cinema. *Differences: A Journal of Feminist Cultural Studies, 14*(3, Fall), 89–111.

Dow, S. P. (2010, December 1). Parallel prototyping leads to better design results, more divergence, and increased self-efficacy. *TOCHI1704–18 ACM-TRANSACTION, 17*(4), 1–24.

Dow, S. P., et al. (1 décembre 2010). Parallel prototyping leads to better design results, more divergence, and increased self-efficacy », *ACM Transactions on Computer-Human Interaction, 17*(4), 1–24. doi:https://doi.org/10.1145/1879831.1879836

Eco, U., & Paci, C. (1983). The scandal of metaphor: Metaphorology and semiotics. *Poetics Today, 4*(2), 217–257.

Eco, U., & Sebeok, T. A. (Éds.). (1988) *The Sign of Three: Dupin, Holmes, Peirce* (1 ed.). Bloomington: Indiana University Press.

Eilouti, B. H. (2009, juillet). Design knowledge recycling using precedent-based analysis and synthesis models, *Design Studies, 30*(4), 340–368. doi:10.1016/j.destud.2009.03.001

Eluard, P. (1966). *Capitale de la Douleur suivi de L'amour la Poésie* (Gallimard French).

Feuillie, N. (2002). *Fluxus dixit. : Volume 1, Une anthologie*. Dijon: Les Presses du réel.

Foucault, M. (1980). *Power/knowledge: Selected Interviews and Other Writings*, 1972–1977, éd. par Colin Gordon, 1st American Ed edition (pp. 194–195). New York: Vintage.

Freud, S. (1997). *The interpretation of dreams*. Ware: Wordsworth Classics of World Literature.

Genette, G. (1979). *Introduction à l'architexte*. Paris: Seuil.

Gentès, A., & Béjean, M. (2011, October). Making sense of constellations of objects: A case study of computer-aided writing practices in theatrical staging. In *Proceedings of the IASDR (International association of societies of design research), 4th conference*, Delft.

Gentès, A., & Cambone, M. (2013). Designing empathy: The role of a "control room" in an e-learning environment. *Interactive Technology and Smart Education, 10*(1), 31–48. https://doi.org/10.1108/17415651311326437.

Gentes, A., Valentin, F., & Brulé, E. (2015). *Mood boards as a tool for the "in-discipline" of design*. In IASDR, https://hal.archives-ouvertes.fr/hal-01246049/

Goldschmidt, G. (1994). On visual design thinking: The vis kids of architecture. *Design Studies, 15*(4), 158–174.

Harrowitz, N. (1984). The Body of the Detective Model: Charles S. Peirce and Edgar Allan Poe. In Eco et Sebeok, *The sign of three* (pp. 179–197).

Hatchuel, A., et Weil, B. (mars 2002). C-K theory: Notions and applications of a unified design theory. In *Proceedings of the Herbert Simon International Conference on « Design Sciences »*.

Hatchuel, A., Weil, B., & Collectif. (2014). *Les nouveaux régimes de la conception : Langages, théories, métiers*. Paris: Editions Hermann.

Hockney, D. (2006). *Secret knowledge: Rediscovering the lost techniques of the old masters*. London: Thames & Hudson.

Holmberg, B. (1995). *Theory and practice of distance education* (2nd ed.). London/New York: Routledge.

Huyghe, P.-D. (2009). *Commencer à deux : Propos sur l'architecture comme méthode*. Paris: Editions MIX.

Ingold, T. (2010). The textility of making. *Cambridge Journal of Economics, 34*(1), 91–102. doi:https://doi.org/10.1093/cje/bep042.

Jacquinot, G. (1993). Apprivoiser la distance et supprimer l'absence? ou les défis de la formation à distance. *Revue Française de Pédagogie, 102*, 55–67.

Jeanneret, Y., & Souchier, E. (1999). Pour une poétique de l'écrit d'écran. *Xoana, 6–7*, 97–107.

Kavakli, M., & Gero, J. S. (2001). Sketching as mental imagery processing. *Design Studies, 22*(7), 347–364.

Kelley, T. (2002). *The art of innovation*. New York: Broadway Business Books.

Klee, P. (1966). *Paul Klee on modern art* (Reprint ed.). London: Faber & Faber.

Knappett, C., & Malafouris, L. (Eds.). (2008). *Material agency: Towards a non-anthropocentric approach*. Berlin: Springer.

Krippendorff, K. (1989) On the essential contexts of artifacts or on the proposition that "design is making sens (of things), *Design Issues, 5*(2), 9–39.

Lakoff, G., & Johnson, M. (1980). *Metaphors we live by*. Chicago: University of Chicago Press.

Latour, B. (1999). *Pandora's hope: Essays on the reality of science studies* (1re éd, p. 47). Cambridge, MA: Harvard University Press.

Latour, B., & Weibel, P. (Eds.). (2005). *Making things public: Atmospheres of democracy* (1st ed.). Cambridge, MA/Karlsruhe: The MIT Press.

Laurel, B. (1993). *Computers as theatre*. Boston: Addison-Wesley Professional.

Leupen, B et al. (1997). *Design and analysis* (1re éd.). New York: Wiley.

Moore, M., & Kearsley, G. (2011). *Distance education: A systems view of online learning*. Belmont: Wadsworth Publishing.

Nelson, H. G., & Stolterman, E. (2012). *The design way: Intentional change in an unpredictable world* (2nd ed.). Cambridge, MA/London: The MIT Press.
Oudard, J.-P. (1969). Cinema and suture. In N. Browne, & J. Hillier, (Eds.), (1989) *Cahiers du Cinema 1969–1972. The politics of representation* (Vol. 3, pp. 45–57). Boston: Harvard University Press.
Oxman, E. R. (1994). Precedents in design: A computational model for the organization of precedent knowledge. *Design Studies, 15*(2), 141–157.
Panofsky, E. (1969). *L'oeuvre d'art et ses significations*. Paris: Gallimard.
Rastier, J. (2002). Enjeux épistémologiques de la linguistique de corpus. In*Journées de Linguistique de Corpus*. available http://www.revue-texto.net/Inedits/Rastier/Rastier_Enjeux.html.
Roozenburg, N. F. M. (1993). On the pattern of reasoning in innovative design. *Design Studies, 14*, 4–18.
Rosenberg, R. (2008). Le schéma de composition, outil et symptôme de la conception du tableau. In La Documentation Française, *Histoire de l'art en France au XIXe siècle* (pp. 419–431). Paris: Roland Retch et al., s. d.
Saper, C. J. (2001). *Networked art* (1st ed.). Minneapolis: Univ Of Minnesota Press.
Sass, L., & Oxman, R. (2006, mai). Materializing design: The implications of rapid prototyping in digital design, *Design Studies, 27*(3), 325–355, doi:https://doi.org/10.1016/j.destud.2005.11.009
Schon, D. A. (1984) *The reflective practitioner: How professionals think in action* (1re éd., p. 79). New York: Basic Books.
Schön, D. (1992, March). Designing as reflective conversation with the materials of a design situation. *Research and Engineering Design, 5*(1, Springer), 3–14.
Schön, D. A., & Wiggins, G. (1992). Kinds of seeing and their functions in designing. *Design Studies, 13*(4), 135–156.
Schurz, G. (2008). Patterns of abduction. *Synthese, 164*, 201–234.
Shank, G. (2001). It's logic in practice, my dear Watson: An imaginary memoir from beyond the grave. *Forum Qualitative Sozialforschung/Forum: Qualitative Social Research, 2*. http://www.qualitative-research.net/index.php/fqs/article/viewArticle/970
Sowa, J. F., & Majumdar, A. K. (2003, Springer). *Analogical reasoning* in International Conference on Conceptual Structures (pp. 16–36). http://link.springer.com/chapter/10.1007/978-3-540-45091-7_2
Stahl, G., Koschmann, T., & Suthers, D. (2006). Computer-supported collaborative learning: an historical perspective. In R. K. Sawyer (Ed.), *Cambridge handbook of the learning sciences* (pp. 409–426). Cambridge: Cambridge University Press.
Tufte, E. R. (2006). *The cognitive style of powerpoint: Pitching out corrupts within* (2e éd.). Cheshire: Graphics Press.
Tulving, E., & Thomson, D. (1973). Encoding specificity and retrieval processes in episodic memory. *Psychological Review, 80*(5), 352–373.
Unwin, S. (2009). *Analysing architecture* (3e éd.). London/New York: Routledge.
Visser, W. (2006). Designing as construction of representations: A dynamic viewpoint in cognitive design research. *Human–Computer Interaction, 21*(1), 103–152.
Vitruve. (1995). *Les 10 livres d'architecture de Vitruve* (2e éd.). et augm (Editions Mardaga).
Wisser, W. (2006). *Designing as construction of representations: A dynamic viewpoint in cognitive design research. Human Computer Interaction, 21*(1), 103–152.

Chapter 6
Design as Debate: The Thing Beyond the Object

6.1 Design as Debate: First Definition

In Chap. 5, I considered design as a field of tensions and introduced composition as a principle of design activity. By composition, I meant the careful gathering and ordering of elements that, through projective abduction, build up new knowledge out of new artifacts. Our examples showed how designers juggle with ideas, materials, tools. But designers also engage with other people, organize collaborations, and gather feedback. They participate in an economy of contribution, the advantages, merits, or limits of which are today largely discussed.[1,2,3] These different facets of collaborative, or participatory, design have been described in the literature.[4] A conference has been dedicated to these subjects since 1990 (Participatory Design Conferences). While the economic advantages of better crafting an artifact by adjusting it to a potential customer through users' participation is pointed out,[5] it is also emphasized that there is a political model at play in co-design. For instance, Luck, who studies dialog in participatory design notes that:

> The democratic principle underpinning participatory design is demonstrated through the involvement of different users during design discussions and through their potential equal contribution to the design outcomes. In this way, the diversity of views expressed by people during the design decision making process can influence the final outcome of a project. The egalitarian, non-discriminatory principles of participatory design are common with an 'inclusive' approach for the design of environments, which should not discriminate on accessibility.[6]

[1] Hippel (2006) and Baldwin et al. (2006).

[2] Prahalad and Ramaswamy (2004).

[3] Zwick et al. (2008).

[4] Sanoff (2000), Cross (1971) and Battarbee and Koskinen (2005).

[5] Ingi Brown.

[6] Luck (2003).

A. Gentes, *The In-Discipline of Design*, Design Research Foundations,
DOI 10.1007/978-3-319-65984-8_6

This chapter studies a certain aspect of participatory design: its social and public dimensions exploring the meaning of artifacts and services. More often than not, technologists are asked to produce better technologies that empower people (though who these people are remains to be discussed). Industries organize contributions to be more efficient, more creative, to insert the objects or services more easily into the fabric of everyday life. Still, the public considers it a right to understand inventions and innovations. The fast pace of technological and societal changes, as pointed out by the philosopher Lyotard,[7] is the focus of concern for societies that do not consider these changes to be part of progress anymore. In the design and HCI communities, this concern is being addressed by such trends as "critical technology"[8][9]or "critical design"[10] or ludic design,[11] that question the values embedded in technologies and that strive to foster a dialog with different stakeholders. The purpose of these trends is not only to organize contributions to increase industrial efficiency nor to make sure that there is a fair representation of the potential users within the industrial process. In the examples we examine, the objects are also a pretext to organize debates around questions of society: death (James Auger, "Afterlife"[12]), spirituality (Bill Gaver, "Prayer companion"[13]), and biotechnologies (Tobie Kerridge, "Biojewellery"[14]). While we understand the social and political need for these debates in general, we also need to understand how designers participate in them while still being designers and not journalists nor politicians. The question is what can design bring to the discussion about sciences, technologies, and societal issues? If design bridges the unknown (inventions) to the known (artifacts and social situations), as pointed out by Le Masson, Weil, and Hatchuel,[15] if it finds ways to relate artifacts and services to people, within specific situations and therefore works on the integration of different worldviews, how can it, at the same time, raise a critical awareness of these emerging technologies? It might seem contradictory that design should be socially integrating technologies while fostering debate about these technologies. Any design object can be discussed. But is it the purpose of design to start a debate? And how does this debate contribute to a conceptive and expansive situation? Our goal is to come up with a better definition of debate but also to start a typology of critical design practices that can be contrasted with other design practices.

This chapter considers several facets of what debate means. More precisely, I want to observe how the object itself is involved and to analyze the way it is discussed in society. If design is indeed correlated to debate, we need to understand

[7] Lyotard (1984).

[8] Agre (1997).

[9] Boehner et al. (2005).

[10] Gaver and Dunne (1999).

[11] Gaver et al. (2004).

[12] http://www.auger-loizeau.com/index.php?id=9

[13] Gaver et al. (2010).

[14] http://www.biojewellery.com/ (accessed in 2015, no longer accessible)

[15] Le Masson et al. (2017).

what constitutes a debating activity in design/conception. I chose to look at three sets of examples that belong to different fields of the design spectrum: art, technical research projects, and eventually design with a particular focus on "critical design". While very different, I would suggest that the critical aspects of these practices shape a territory of design and politics.

To start this discussion, it seemed interesting to study a work of art that triggered a number of debates in France in 2001. "Tell me your secrets" is exemplary in two respects at least. First, it was censored. Second, the artist arranged what we could call a "debating ring", using the internet as a platform to share information and provide a forum for opinions. Studying this case seemed like a good way to understand the aesthetics of controversies.

The second example is about design in research projects. Design in that case is often used to make an invention more palatable to the bigger audience. At the time of the emergence of the technical object, social scientists and designers organize feedback with a larger circle of users to test a prototype and adjust it to the public. This situation of test and adjustment is also a situation of communication, where participants get a chance to express their frustration, their expectations, their disagreement. In effect, the debates are not necessarily very heated, but they steer the technical object towards unexpected developments, new ideas, and new solutions.

Thirdly, a number of designers claim that they are engaged in "critical design". They want to question the social adequacy of certain technologies or social evolutions before we become fully engaged and committed to them. They consider that design – being part of the industrial system – can and should engage in the production of new artifacts and systems not necessarily in a more responsible way, but at least in a concerted way. They consider that they have special "designerly" ways to introduce a debate about science and technology. Starting with "uncanny" objects, they organize their activity around the "showroom".[16] This indicates their determination to play a role in the evolution of values: inciting debates is a new goal for design.

These three fields: art, research, critical design, have their own distinct approaches to debates and artifacts, but, in this chapter, I also consider what they may have in common.

6.2 Nicolas Frespech: "Tell Me Your Secrets". A Story of Censorship in Art

As I was working with artists producing works of art specifically for the internet,[17] I got interested in Nicolas Frespech and we finally met and discussed his production that included very strange pieces like "La maison des immondes pourceaux" ("The

[16] Koskinen et al. (2011).

[17] Gentes (2001).

House of the Filthy Swines"). At approximately the same time the "affair" of the "Secrets" started.

Nicolas Frespech's "Tell me your secrets" was a simple interface that allowed people to write a few words (a secret) and post it on the internet. In December 2001, 3 years after its acquisition by a regional contemporary art fund in France (FRAC Languedoc Roussillon), the art work "I am your friend. Tell me your secrets" disappeared purely and simply from the internet. Nicolas Frespech only realized this after a few days. He first thought that there was a breakdown in the server and enquired about it with the FRAC. He did not get a very clear answer and therefore turned towards the company Zarcrom that ran the servers for the FRAC. Zarcrom did not respond either. Nicolas then realized that the work of art must have been removed by the FRAC itself. The affair of the "Secrets" had just started.

Why was this work withdrawn from the Internet? And what was the artist going to do about it? As it happened, Nicolas started a new socio-poetic experiment. My investigation, done with the help of members of the National Digital Art Fund of the French Ministry of Culture, made it clear that the withdrawal of "Tell me your secrets" started a new artistic intervention from Frespech, who then exploited the plasticity of the internet to build a debate on privacy, art, and public institutions.

6.2.1 *Public Art and Private Contributions: Aesthetics of the Internet*

"I'm your friend: Tell me your secrets" (1998) was a contributory artwork. People could leave a secret of a line or two, which was then distributed on the internet. Everyone could then read the succession of "confidences" in the center of a pink screen. This simple invitation garnered rather personal expressions on sex, other people or professional frustrations, etc.

The users wrote without *a priori* censorship[18] and were preserved by anonymity. The intimate space of writing was impersonally organized as a form with a series of boxes to be filled.[19] Long before Facebook or Anonymous, "Tell me your secrets" started a reflection about expression and anonymity on the net. At the time, I qualified this as "ex-timacy",[20] i.e. an intimacy that was broadcasted and staged for an

[18] In fact, the messages were screened by the author before being posted to eliminate any racist or pedophile messages.

[19] The artist had started the whole project by receiving emails and posting them himself, but finally received too many of them and developed an application so that people could directly submit their message.

[20] Gentes (2007).

audience and at the same time unreachable. While the real intimacy escaped so to speak in the background, there was but very little that was left to the public: a disembodied rather meaningless accumulation of sentences.

In fact, Nicolas Frespech illustrates perfectly what Virilio quoted by Stallbrass qualifies as social cybernetics:

> Paul Virilio has argued that the internet is not an example of liberty but of social cybernetics. In other words, humans act as the feedback mechanism in a system that has its own autonomy: in visiting and clicking through sites, atomized users provide data about themselves that guides the machine to perfect itself; the machine evolves not necessarily to serve users' need but to exploit the dataset that is the sum of their inputs. Exercising the freedom to choose between limited and discreet options, they feed the system.[21]

Though people might think that "Tell me your secrets" would be a treasure trove of original experiences and statements, the artist admitted[22] that he actually had to remove a lot of the sentences, not because he thought they were inappropriate, (they were rather innocuous in particular in the context of contemporary art), but because many were strictly identical and created a long and very boring series. It actually confirmed that "the offer of participation, as in computer games with their rigid constraints and criteria for success, may merely intensify conformity".[23]

However, it is not how the local powers saw it. The Secrets were judged shocking by the regional Court des Comptes (Regional Treasury). It was:

> noted without coming to a conclusion about its artistic quality, that this work, which had been bought with public funds, comprised obscenities which could only run up against the sensitivity of a public not informed.[24]

The contributions were considered too indecent for the region by authorities who did not want to somehow endorse them. It is to be noted that the regional funds for contemporary Art (FRAC) are very much dependent on local authorities and are subjected to political changes more than other public institutions. At that time, the elections in the region brought to power an extreme right party that favored middle of the road productions. Subsequently, the website was removed from the internet.

6.2.2 *Offering a Space for Debate*

Following these events (2002), Nicolas Frespech made an incredible turnaround. He started a debate on his website by providing news, archives, comments, links towards all the actors concerned. But his website was only a small part of his

---

[21] Stallabrass (2003).

[22] In an interview with the author.

[23] Stallabrass (2003).

[24] These reasons are published and discussed in a number of press articles: *Les Inrockuptibles, Digipress, Libération.* See Bertrand Gauguet, « Les secrets censurés de Nicolas Frespech ou comment *Je ne suis plus un site* », in *Archée*, avril 2002, http://archee.qc.ca. Voir aussi le récapitulatif de l'histoire de cette œuvre sur: http://www.20six.fr/lessecrets/

strategy. He also used all the media resources of the Internet to talk about censorship and his work. Thus, began a second life for the "Secrets", which specifically played on intermediality. "I am not your friend" – as the introductory sentence to the "Secrets" was modified – became a public space of a different nature in response to a bureaucratic situation, defined by governmental power. To fully experience this particular work of art required following the links and exploring the entire canvas woven around the missing work.

If one checks on the website of the FRAC, Nicolas Frespech is present in several fields, mainly in the archives as well as in a list of artists. But the works that are cited are "The Filthy Swine," "And I and I" and "Picnic of Art" (1997). The work "The Secret" has indeed disappeared from the site. It is not even mentioned in a list of acquisitions. At that time (2001–2003) a Google search with the words "secret" and "Frespech", gave very diverse results on the destiny of the "Secrets". A computer magazine came up with a debate on the issue of pornography.[25] An online artistic magazine – Archee – introduced an artistic analysis.[26] This analysis was then quoted by another website that was more specialized in activism on the net: Bugbrother.[27] There were interviews in a Québécois journal: Le National.[28] The artist sent emails published by webzines like Visual Image or Uzine.[29] On Wikipedia, Nicolas Frespech's students posted an article.[30] The festival Art Outsider devoted a year to studying censorship where the « Secrets » occupied a major place.[31] Nicolas Frespech's "Secrets" is also mentioned in the website of another artist, Antonio Muntadas and « The File Room », that archives censored works of art under the guise of technical and legal files:

http://www.thefileroom.org/documents/dyn/DisplayCase.cfm/id/384

Name: I'm your friend...tell me your secrets. Je suis ton ami(e)...tu peux me dire tes secrets.
Date: 1995–2005
Location: Europe

---

[25] http://www.advancedinformatique.com/article.php/id/29

[26] http://www.archee.qc.ca/ar.php?page=article&no=181

[27] http://www.bugbrother.com/article185.html

[28] http://www.le-national.com/frespech022002.html

[29] http://www.visuelimage.com/ch/frespech/index.htm
Je ne suis plus un site/ I am no longer a website
http://www.uzine.net/breve656.html
"Net art ta gueule
Jeudi 20 décembre 2001. Même situation au 10 janvier 2002
« ... Ce site est. actuellement "prêté" et présenté sur le site de l'Ecole nationale des Beaux Arts de Lyon, dans le cadre de l'exposition "dévoler" qui a été organisée cet été.
http://enbalyon.free.fr/frespech/index.html
J'aimerais donc que vous m'aidiez à comprendre les vraies raisons de cette nouvelle forme de censure. Espérant que cette mauvaise expérience pourrait nous permettre de réfléchir sur les enjeux de l'art en ligne, du politique dans les choix artistiques, et espérer aussi un nouveau statut pour les "cyberéalisations".
Vous pouvez me contacter à cet E.mail: immonde@cicv.fr

[30] http://fr.wikipedia.org/wiki/Nicolas_Frespech

[31] http://www.art-outsiders.com/archives4/default.htm

Subject: Political/Economic/Social Opinion
Medium: Electronic Media
Artist: Nicolas Frespech. Frac Languedoc-Roussillon. France
Confronting Bodies: Nicolas Frespech VS Conseil Régional Languedoc-Roussillon
Date of Action: 7 december 2001
Specific Location: Montpellier (France)
Description of Artwork: A collaborative artwork on the net presenting secrets, sent by e.mail.
Description of Incident: The 7 december, the http://www.fraclr.org/secret was illegally closed by Zarcrom, the host of this artwork. The URL was closed.
Results of Incident: No body can have access to this art work, and me too, i can take this web elements, the ftp was closed too.
Source: Artist source: Nicolas Frespech
Submitted By: Frespech Nicolas.

There were texts about the "Secrets" everywhere on the internet, in websites with various status and vocations. Not only did the artist activate all "the network of people whose cooperative activity, organized via their joint knowledge of conventional means of doing things, produce(s) the kind of art works that art world is noted for".[32] He put the whole art world network on stage. Frespech orchestrated the censorship experience of the "Secrets" by showing the dissemination and processing of information through various channels on the net. This orchestration became a work of art in its own right. As Craig J. Saper, defining Networked Art, contends:

> When aesthetic and poetic decisions embodied in artworks lead to a heightened or changed social situation, one needs to describe these forms as sociopoetic rather than as artworks within particular social contexts. The social situation is part of a sociopoetic experiment.[33]

Craig Saper dates these networks experiments back to the fifties when artists like Arman, Piero Manzoni, Martial Raysse, started playing not only with assemblage but also with the mail itself, in particular Yves Klein who painted a Blue Stamp on an envelope that was taken for a real one and used by the Post Office. Like his predecessors, Frespech used the workings of a social and technical system to create a socio-poetic experiment. The media was not simply a communication channel but was actually foregrounded so that it transformed and engaged the spectator to look at the way Internet worked. "The term sociopoetic describes artworks that use social situations or social networks as a canvas".[34]

6.2.3 *Lessons from Nicolas Frespech's "Tell Me Your Secrets"*

Governments censoring communication are common. "Tell me your secrets" was special in that it used network technology to successfully spur a public aesthetic/political conversation.

[32] Becker (1982).
[33] Saper (2001).
[34] Saper (2001).

The first lessons are formal and semio-pragmatic lessons.

The purpose of this example is to understand first how a debate can be embedded within an object. What are the seeds for debate within the object? Even though we are interested in the formal qualities of these objects, we need to be aware that they challenge preconceptions, at a certain time and a certain place. In other words, the same objects might not be so controversial in another society or at a different period of time.[35] The first question therefore requires a semio-pragmatic analysis of the objects. How does this object make sense in a certain situation?[36] The first lesson of our research was therefore a lesson in methodology. The study of contents has to be studied in context, which means that one needs to connect the discussions to the artifacts.

Nicolas Frespech's "Tell me your secrets" obviously was challenging the sense of propriety of the public buyers. The "obscenity" of the "Secrets" shocked. Nicolas Thély, art critic and Professor of aesthetics at the University of Rennes, remarked that the free access to the work of art was also a critical aspect of the question: the fact that it could be accessed in other places than a museum or an art gallery and could in effect be seen everywhere implied that the usual cultural framework that helps people understand the meaning of art was absent. Therefore, it was controversial not only because its "contents" were deemed provocative but also because the situation of communication – that generally frames potentially "polemical" contents – could not be guaranteed. "Tell me your secrets" was challenging curating principles[37] because there was no cultural mediation. Obscenity is defined not only by the contents but also by the mediation.

This example also gave me a communication lesson.

From there, the orchestration of a debate meant that different actors discussed the now absent "Secrets", its artistic worth, and what it revealed about our society. The artist distributed the information on the internet, was interviewed by journalists who also published and discussed the information. People read about it and could form an opinion. The second stage of the "Secrets" is the operational stage of the debate. This operational stage is about the use and organization of a public space for communication. It implies the identification of actors to debate with, the selection of the proper channels of communication to reach the expected audiences, the crafting of an argument and the texts supporting this argument, as well as the genre and style of media and messages. In this particular debate, the object is expanded beyond its tangible existence (or non-existence) as the discussions involve a reflection on art in our society, on the internet, on contribution, etc. In doing so, the actors actually augmented the meaning of the object beyond its formal characteristics, by associating it with a number of aesthetic, social and political questions – for instance who can talk to whom, with what kind of messages – questions that define the modalities of governance in society.

---

[35] Grenier (2008).

[36] Hymes (1974).

[37] Gentes (2003).

Starting a debate with an object therefore means at least two things:

- a choice of contents that challenges what some groups consider art in terms of propriety of subject and representation as well as proper space for art.
- a situation of communication that relies on other media and organizes a viral circulation of the information. But the situation of communication is not only defensive or rhetorical. It is also meaning-making: it creates new meaning for the artifact.

Since the Avant-garde, this exploration of how to circulate, distribute, comment art, outside proper art spaces, has been a current occurrence. However, is there anything remotely similar in science? As I turned to engineering research projects, I was actually surprised to recognize some similarity in research: the object – the demonstrator – is somehow designed to trigger a debate – during the evaluation. I therefore looked carefully at the organization of this specific situation.

6.3 Debate in Research: Designing the Demonstrator

A research project hardly seems to be the place for an open debate about its own value. When studying the narratives in Chap. 4, we have seen that they are rather supportive of the technology. Researchers will defend their own work and do not over welcome criticism. Others should therefore be in charge of this critical discourse and are: political institutions, the press, through popularizing debates. But a closer look at how research is done shows that engineers and designers do bring some debating features into the research process. In fact, there are elements that *de facto* organize the way the project is going to be transmitted and discussed. In particular, the applied dimension of the research, the fact that in some projects, the protagonists need to consider social and economic perspectives, bring some elements of debate into research. This section focuses on how this debate is introduced. It is my hypothesis that prototypes are not limited to testing the technical part but play a political and expansive role.

In this section, I want to look at the "demonstrators" of three research projects about mobility and pervasive computing that I already presented in the course of this book. A demonstrator is a prototype made by the design team for testing purposes. The projects developed new systems based on wireless ad hoc networks (ANR-Safari and ANR-Transhumance) or RFID (ANR-Plug) network, and tested the technical and social potential of these technologies. Several applications were developed in this context:

- A fighting game in the Montparnasse station (SAFARI) in which the virtual world uses the same architecture as the station, and fights begin and end with the arrival and departure of real trains.
- A quest game in the neighborhood of the Butte aux Cailles in Paris (TRANSHUMANCE), where players must find Street Art on the walls, and pool their discovery.

- Finally, an educational game in the Museum of Art and Crafts (Musée des Arts et Métiers) (PLUG) in Paris, where players must collect images that they have to properly attribute to objects, and exchange with other players with their phone in the museum.

Comparing the three cases, I could see that some formal elements directed the discussions. In particular, two aspects kept coming up:

- Objects did not look like ready-for-market products (though sometimes people did want to buy them). They were "open", that is, they presented themselves as unfinished. They had what I suggest to call "infra-design" properties that let people know that they were to develop and to complete the artifacts.
- Because researchers wanted feedback, they also built their demonstrators with a "transparent" design that let users understand what is specific of the technology. The artifacts had somehow "see-through" characteristics.

But the situation and communication elements were also designed to organize the discussion:

- The cultural mediation emphasized a specific time orientation: the tests were devoted to examine "objects to be". They were oriented towards the future.
- The space where the experimentation took place was designed as an experimental field that, consequently, opened the possibility of transgressing some of the social rules generally associated with the location.
- Finally, the experiment was situated in a specific institutional framework: it was controlled by a research establishment.

All these elements structured the debates as shown in the following sections. But first, I want to point out the epistemological complexity of the object. The demonstrator is the "moment of truth" on many levels. Actors comment that it is not only a scientific achievement. It is also an artifact that represents the group and its capacity to work collaboratively. Moreover, the final result should show some aesthetic merit. The implementation of the demonstrator is therefore always a stressful time because so much is at stake.

6.3.1 The Epistemological Complexity of the Demonstrator

The demonstrator is a technical object which evolves during the project around a quest for coherence. Its "concretization"[38] is based on the architecture and final integration of the different technical parts, some of which are original contributions, others are existing mature technologies.

Convergence is critical because the technology is defined by its different modules making a system: it is an epistemological claim. The group coordinators are under a lot of pressure to ensure that the modules effectively work together:

[38] Simondon (2001–2016).

> Hello everyone,
> We have well advanced developments in the Device Manager module, but the following point is now we have a problem: we have not decided on the **mode of communication between the modules of the platform.**[39] (in bold in the text)

Many interactions deal with fitting all the modules together. The success is partial if one or two modules (that still work by themselves) cannot be included. The validity of the project depends on its achievement as a system and not only as independent parts. This means that brand new findings can be merged with more mature technologies even obsolete ones!

> Hello everyone,
> FYI, the new N800, the new internet tablet from Nokia.
> http://www.clubic.com/actualite-67883-n800-nokia-tablette-internet.html
> And here we are on obsolete hardware! ;-)
> Cheers![40]

Sometimes, researchers may have to justify the results from a range of social as well as technical criteria. As we have seen in Chap. 3, social scientists are then asked to address the question of usability and user acceptance and therefore often participate in the assessment and validation of the technique. Technical elements are brought and tested together but also are merged with social and cultural considerations that expand the technical concepts beyond their first scientific reach. The demonstrator therefore also embeds different models of what research is, for instance, feasibility over breakthrough, or social hypothesis over technical prowess. Each subgroup of the project has "partial jurisdiction over the resources that the object represents".[41] Researchers then confront the difficulties associated with project management. As actors' priorities do not necessarily coincide, the choice of technologies, protocols, devices can be challenged during the project. The demonstrator always shows traces of the different tensions, disagreements that were not resolved during the project: "this resolution does not mean consensus. Rather, representations, or inscriptions, contain at every stage the traces of multiple viewpoints, translations and incomplete battles".[42]

The interviews that I had with the research partners showed that they can strongly resent these imperfections. When comparing projects, some of the actors mentioned their appreciation of their tightly knit group because, amongst other reasons, "it shows" in the final results and in particular in the demonstrator.

[39] Transhumance, January 2007.

[40] Project ANR Transhumance, mail sent to the research team who works on the version n 770 of a Nokia tablet, January 2007: « Bonjour à tous, Pour info, des nouvelles du N800, la nouvelle tablette internet de Nokia.http://www.clubic.com/actualite-67883-n800-nokia-tablette-internet.html Et voilà, on est. sur du matériel obsolète!;-) A bientôt ».

[41] Star and Griesemer (1989) p. 412.

[42] Ibid., p. 413.

> There was a huge difference between Transhumance and Popeye. There was hardly ever any real convergence and aggregation in Popeye. It is a question of management. The group was too big. It is impossible to get everybody to really work together. […] The demonstrator is not as interesting.[43]

Tensions are considered a problem that the team makes an effort to overcome not only for operational reasons but also for aesthetic reasons. The demonstrator should be good-looking. The poetics of technological and social identity are at stake. The team wants to build something beautiful and complex that reflects their own harmony.

To sum up, every effort is being made, within the constraints of the project, so that demonstrators enticingly show off the technology to the different stakeholders. As part of the system of proof, demonstrators have to present reproducible results. As part of the system of communication, they have to display understandable and hopefully attractive results.[44] The demonstrator therefore appears as a way to articulate technological elements but also to connect with people and potential activities and to be a representative of the research group. It has therefore been analyzed as an asset in a negotiation to get more funding and secure future research or entrepreneurial developments. As pointed out by Smith, "IT demonstrations are regarded as an essential part of making adoption decisions".[45] The launch of the technology means the accomplishment of a program but also sets off negotiations for future research programs or products.[46] However, before being part of a financial and organizational negotiation, I contend that the experimental phase of the demonstrator bolsters debates of opinions that are supported first by the design of the devices and second by the situation of tests. Debates are present within the objects and through the situation of communication that opens the process of interpretation, of possible associations, of debatable choices.

6.3.2 *A Feeling of Closure*

The demonstrator results from a number of choices that are made during the projects and that *de facto* limit the potential scope of the emerging technology. When the project is almost over, the team has accomplished a difficult task: it has temporarily fixed the identity of the invention. But for the engineers and the designers who collaborate on these projects, the expected outcomes are quite different. I have partially described the different narratives that project the invention into its potential world, from an engineering and from a design points of view. I want to come back to these narratives as they are embedded in the demonstrator.

---

[43] Interview of one of the researchers involved in both projects, 2008.

[44] Smith (2009).

[45] Smith (2009).

[46] Simakova (2010).

The process of identity building is double sided: it is a process that brings out the originality of the project and, at the same time, it is a process that connects the innovation to other existing practices or artifacts. From the engineering point of view, what is at stake is the coherence of parts that are developed more or less independently in the first place. What Simondon describes as concretization is at play--to finely tune the different elements together. For the designer, there is no technical determinism: he/she is not going to manifest the essence of the technology and he/she does not fulfill a "need" that so far no one even knew existed: "Designing as disclosure can be presented as a practical resistance to such naturalized ways of seeing, thinking and acting".[47] The main task of the designer in this context is therefore to give some stability to the technology by giving it a tangible form, enrolling it in a family, and staging it in a situation.

Giving a shape of course limits the range of options. But the implementation is not only a decision reducing the possible scenarios. Devices, shapes, colors, aesthetics, points to new possible interactions and understandings. In other words, from a designer's point of view, the materialization is a way to open to new interpretations that change from one public to another, in particular because the object is always situated, hence influenced, by its contexts. The research of a family of similar objects is also a way to build an identity by comparison as we have seen in Chap. 5.

> Categories of designs are often described by showing groups of images that illustrate the range of designs that belong within the category [...] What makes the images similar may not be obvious and may not be explained. Often one or two images are highlighted, for example by making them bigger, because they either show extremes within the space or provide typical examples.[48]

Objects that look alike, bring with them a certain form of mimetic understanding of use, gesture, context. In their contribution to the SAFARI Research project on distributed mobile networks, the designer Jean-Louis Fréchin and his students explored the common territory of similar devices (Fig. 6.1).

It was a way to define the communicative ecology of the object. The collage of these different pictures of artifacts on the same slide foregrounded their commonalities. But at the same time, they opened to a diversity of options on features, shapes, appearances that left open the idea that the team could actually choose another model.

Eventually, designers rely on their culture of objects in situation and put forward the need to relate the device with its physical environment. Architecture, lighting, commercial boards, will influence the new artifact and the latter will influence its environment. The issue is no longer limited to the device and its handling by a user but encompasses the whole context. Pervasive computing was for instance considered as a global question that could impact the layout of the site, the activities, the use of other media. This consideration led the designers to think carefully about the

---

[47] Newton (2004).

[48] Eckert and Stacey (2000).

Fig. 6.1 "Devices" presenting the family of wireless products as inspiration for the SAFARI demonstrator, Jean-Louis Fréchin – No design – RNRT – SAFARI

aesthetic and pragmatic interdependence of things, people, and location that they then represented in their presentations (Fig. 6.2).

6.3.3 *Nothing Is Forever: A Very Special Design, "Infra Design"*

However, the demonstrator is a very frustrating experience. The results are not necessarily considered complete or even solid. The demonstrator is often a shaky work in progress event considered flimsy.[49]

For designers, the demonstrator is a source of frustration because it stops the design process before they can actually develop it into a more developed artifact. The first productions can only be considered as "drafts", sketches, even though the demonstrator takes place at the end of the research effort. In short, the conceptual work is rich but the interfaces are poor. It might not even be a proper prototype

[49] Smith (2009).

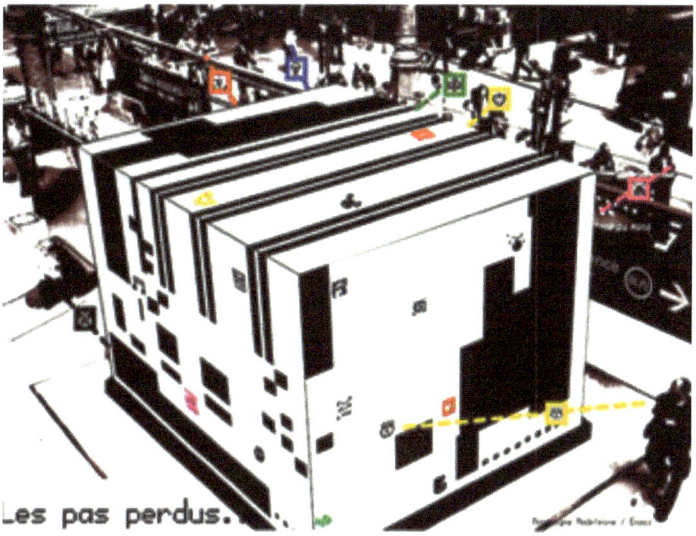

Fig. 6.2 ENSCI Students project: LOST AND FOUND – Ad Hoc project – RNRT – SAFARI

including the main technical characteristics and performances of the new product. For that reason, the engineering team also is not entirely satisfied.

So, what is expected from the demonstrator?

In his article on design as disclosure, Sidney Newton defines design as a projective exercise starting from the reading of a situation / problem. He proposes the concept of design as "disclosure" not as the discovery of a hidden truth but as an iterative and projective process of understanding. "This is why in design it is always difficult to know where to begin and how to end".[50] In my opinion, the demonstrator is part of this projective activity in a very special way because it claims its unfinished status. Demonstrators are not only showing the technology and its potential application. They also show that the research is in progress, that the object needs to be evaluated not only as it is but as it could be. The artifact and the situation where it is displayed and performed are telling those who test it to continue the creative process.

In two of these three cases, the choice of a game for the "demonstrator" appeared as a fruitful option. Not only did the team find it "nicer", "more fun", more "relaxing", less "serious", than working on a "serious" work oriented project. But thanks to this choice they could explore more freely the potential of a technology by inventing rules that followed the potential cues given by the technical properties and the characteristics of the situation.

The table below sums up how demonstrators follow two sets of goals: concretization and conception and how they are oriented at the same time towards a closure of the research and a projection in the future. It is the double nature of the demon-

[50] Newton (2004).

strator that makes it a fundamental element of innovation in engineering and design research.

First, while it fixes new knowledge related to the properties of the technology and the aggregation of different technical modules, it also takes the risk of being defined by an exterior situation that cannot be controlled. Conception sciences are bound not only to theory but also practice because they open up to a new situation that they caused.

Second, demonstrators are not only boundary objects that bring together the different contributions that concretize the new technology. They are generative objects, that lead to other objects, concepts, and methods. A generative object is always transient. In a way, it fights the concretization process described by Simondon.[51] Its purpose is therefore not only to be fixed and to gather correlated knowledge, but to leave an opening to other projects. In this way, it follows an "infra design", that is to say a design that foregrounds its unfinished state. To do so, it is also a transparent design as the following section will argue (Fig. 6.3).

6.3.4 A "Transparent" Design of "Open Objects"

Our goal was to assess inventions and see how people reacted and understood these new artifacts. But more importantly, we needed to have them expand their potential uses. While many options were possible, we finally realized that the best results were when the design of the interface showed the technology through a semiotization of its characteristics on the interface. We figured this out the first time because of a bug. As I described in Chap. 2, we were experimenting with SAFARI in the railway station, the network architecture and status, theoretically "hidden" in the PDA, were unexpectedly accessible to our testers. They immediately started to ask questions and showed a keen interest in what it meant. Though the team primarily

Demonstrator: infra design		
	Concretization	Conception
Object mode	Closure of the internal logic of the object: theory What works / does not work	Opening on the external logic of the object: Practice Does it/ would it fit into a situation
Status of the experiment	Evaluation	Projection
Managerial Purpose	Boundary object	Maieutic/ generative object

Fig. 6.3 General view of the properties of demonstrators

[51] Simondon (2001–2016).

thought it was none of the users' business, they had to recognize that participants with access to some of the technical properties became more involved in understanding and handling the demonstrator. Because they could understand and visualize" technical issues, "testers" projected new proposals for services.

Giving visual cues to the testers on how the demonstrator works therefore seems like a good option if one wants to start a debate about this aspect of the experience. But it is not enough to start a conversation with the participants in the experiments. This "transparency" of the demonstrator is what makes it debatable. But how can we define openness so that it is recognized and reproducible?

The artifact has to show that there is room for change and transformation. It is our hypothesis that demonstrators show their potential for change because they exhibit that they do not work… yet. Testers are confronted with "things" that are not robust, nor reliable, nor efficient, and this is exactly why they feel they can add something to them.

However, it can be a very frustrating and complicated experience because testers are used to perfectly polished products that they find ready "on the shelf". Disappointed expectations have to be turned into a profitable debate if the object is to be continued. The situation of reception is therefore geared so that people have to complete the artifact through their suggestions, critiques, projections. The research framework organizes the difficult transition from a very limited interaction with a device to a powerful space of contributions as we will discuss in the next section.

6.3.5 The Demonstrator in Action: Tests and Interpretation

Tests not only verify functionalities, they also bring the technology into the world. The experimentation is a very special moment as the technology is borne into this world and progressively discovered and connected to different places and people. These first steps need a carefully planned and organized environment that benefits from what Joelle Le Marec calls the benevolence of the actors not necessarily towards the object itself but towards the situation of experimentation and towards research and science in general.[52]

The recruitment is a key aspect of debate in research. In fact, tests are organized according to a gradation in user's skills and expected types of interpretations and feedbacks. The first "batch of testers" is recruited from the research team. Their goal is to ensure the reliability of the device before handing it to people outside the project. This first phase of testing identifies and solves technical bugs. These first

[52] Configurations, 2003, dossier « l'ethos scientifique : autorité, auctorialité et confiance dans les sciences », Le Marec, J., Babou, I., 2003, « De l'étude des usages à une théorie des « composites » : objets, relations et normes en bibliothèque », in Souchier, E., Jeanneret, Y., Le Marec, J., (dir.), 2003, Lire, écrire, récrire – objets, signes et pratiques des médias informatisés, Paris, BPI – Centre Pompidou, pp. 233–299.

testers are very aware of their responsibility towards the accomplishment of the projects and do not judge but improve the technical functioning of the device.

The second phase involves testers who did not design the technical platform and service, but who share a similar professional background with the research team (colleagues and students from the same field, met during a conference for example, or coming from the same campus). They are expected to flush out technical errors that savvy designers from the team would more or less consciously avoid. The logic of these tests is still very focused on reproducibility and the robustness of the device. These recruits have all in common, if not the same knowledge base, a shared experience of research and prototypes. This *a priori* gives them a clear role in helping the team figure out what does not work.

At a later stage of the project evaluation, users are recruited and are observed so that criteria largely used in the field of HCI (Human computer interface) like good technical performance, speed of use, low error, etc. can be appraised. The goal of these tests is to organize a form of closure of the project by providing results on the "functionality" of the technical device.

The last circle of testers is recruited on a larger basis: for instance, for Transhumance, the "Human Resource" application was tested in the offices and with the staff of France Telecom. While technically skilled testers can use their expertise in a way that palliate the eventual defaults of the system, the "novice" testers surprise the research teams by unexpected moves and interpretations of the interface. They also are out in the field, which means that the devices have to resist the field trial and aspects of the situation (including light, architecture, other people, etc.)

As we were going through these various phases of testing, we realized that some issues were not addressed through this process. For instance, each research project usually builds on previous projects and also anticipates what can be developed in the next. In other words, the span of the research is not limited to the duration of the project. While the test phase marks, without a doubt, a closing stage in the research, it can also generate new questions. But "traditional" user tests – as described above – do not meet this logic of "a work in progress". To take this aspect into consideration, we proposed a new form of assessment based on a fourth generation of participants and more importantly on an emphasis on the research context and its goals. I call this type of test: "creative evaluation". I want here to emphasize just two aspects of these evaluations.

- First, they rely on bringing the research process with the actual research team to the foreground.
- Second, they put an emphasis on the contribution of testers to the research program.

This "creative evaluation", that we developed at the Codesign Lab with Camille Jutant et Aude Guyot, was prepared with the rest of the research team. We participated in the meetings and organization of the projects to gather data about the way people build, argue about, and dream of the project. We also interviewed project members individually, to explore with them their ideas about the potential of the technology, their imaginary projections, their hopes, and the difficulties encountered during the project. In other words, we got a better image of the research as a place for desire and imagination, as well as for scientific procedures. This paved the way to how the experiment would be presented to the testers. We then shared some data like the difficulty in describing what the technology is for, the difficulty in defining preset goals, etc. It gave the testers some research background, based on what the researchers expressed, hence introducing a form of dialog.

Eventually, during the evaluation, the participants were introduced to the research questions, and given a detailed explanation about the technical challenges. This meta-communication about the project is important in many ways. First for practical reasons, we need testers to be patient. Using a demonstrator is not the same as using a fully developed product, and a certain kind of "suspension of disbelief[53]" is important to the research project. Then it explicitly frames the participation of the users so that they feel they can contribute to the research in their own terms. The mediation emphasizes the fact that it is an open-ended artifact that needs help to grow – help that can come in many different ways – but also that it is an open-ended situation because the outcomes are not predefined.

In other words, the debate starts because we do not ask to validate or invalidate a proposition but because the whole presentation is about research, and what it means to be doing research in engineering. For instance, we say things like: "research is about something that does not work. If it works, it is no longer research". The experiment is then carried out in the field with observers following participants and asking them to verbalize what they do. Finally, at the end of the experiment, testers comment on the whole experience and often relate it to previous ones, that we could or should relate our research to. One striking thing is the efforts that participants put into not judging the prototype, but into augmenting it. For one of the projects (Transhumance) we could do a comparative study of two types of test: a traditional psycho-ergonomic test versus a "creative evaluation". In both cases, we got the same amount of criticisms of the device but with the creative evaluation, we also got new ideas on how to redeploy the experience in different situations, with different actors, for different services. The emphasis that we put on the vulnerability of research, hence its creativity, structured the interactions in a more expansive way.

6.3.6 Lessons from Demonstrators: An Aesthetic of Bugs and a Crisis Communication

Thanks to this new type of evaluation, we learnt some lessons first about the formal and semio-pragmatic properties of generative demonstrators that finally rely on an aesthetic of "inviting bugs".

[53] Coleridge (1985).

In research, the objects cannot be considered as improper but they are definitely unfinished and they actually foreground their "unfinishedness". This triggers a tsunami of complaints:

– the devices are criticized:

M.	"The surface is less reactive than the Palm. I have to type several times to enter the information".
P.	"I cannot validate the image. The image is good but it does not want it. I think it is a bug"
I.	"the image is not validated. I'm too fast for it".
M.	"we can't read anything because of the sun".[54]
–	Interfaces are stupid:
MB	"we should have a final map. It is the final map that is missing"
M.	"to validate an image.... It is not a task. It does not bring anything to the game"[55]

The list of criticisms is always so long that most researchers would probably want to jump under a train if they were left alone with it. For them, as for the testers, the situation would probably be desperate and meaningless if the communication and mediation using the demonstrator did not reframe the interactions and basically changed the meaning of "bugs". Rather than being considered and presented as problems to be fixed, the situation consists of giving a new value to bugs: a chance to redefine the objet, the situation, the goal of the invention.

We therefore learnt that changing the meaning of bugs was extremely important to the design process and that considering them as "inviting bugs" was part of a communication strategy that turns problems into opportunities. In other words, scientific communication does not hide bugs but turn them into an invitation for testers to contribute or challenge the purpose of the invention. The unfinished, open nature of the artifact is a condition to start a debate.

Looking back at the different experiments and documents, there is actually a gap between how the system is represented in these documents, and the actual objects that participants have to manipulate during the tests (Figs. 6.4, 6.5, and 6.6). Quite a few documents are meant to promote the project and try to hide the "poverty" of the artifact. I think it shows a tension between two different rationales: on the one hand, the promotion of an object that is presented through the elegance of its forms for legitimation purposes, and on the other hand, the unfinished state that is needed to start a conversation. When we look at the actual situation of test, there is a balance

[54] Transhumance, interview with M. « La surface est. moins sensible que le Palm, il faut taper plusieurs fois pour entrer l'info » interview with P. « l'image était bonne mais il ne la voulait pas », interview with MB. « L'image n'apparaît pas comme validée si je suis très rapide ». interview with M. "je ne peux rien lire à cause du soleil"

[55] Transhumance, interview with M « Valider une image. Mais ce n'est. pas une tâche en soi, elle n'apporte rien au jeu ». MB« ce qui me manque c'est. la carte finale »

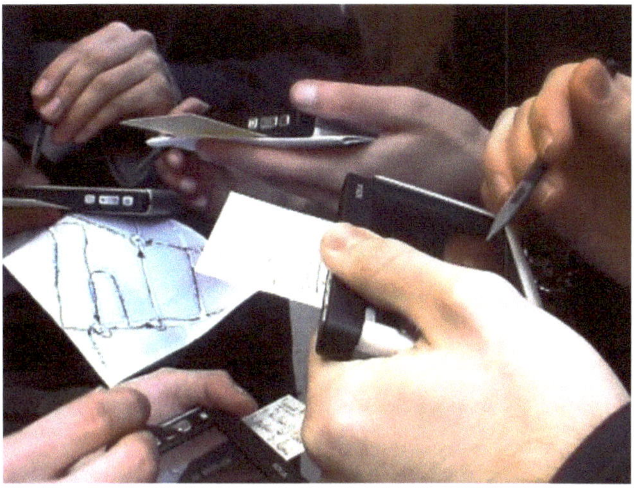

Fig. 6.4 Transhumance tests – The users need maps, paper, plus their device – RNRT – Transhumance, 2005

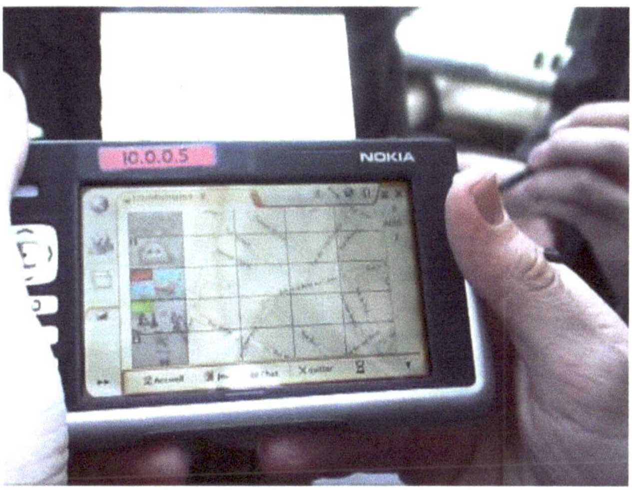

Fig. 6.5 Interface labelled for the team to know what device is working or not-RNRT Transhumance, 2005

that most tests try to reach between a partly dysfunctional object and a situation of communication that we should qualify as a crisis communication strategy.

– First, as in all crisis communication strategies, the crisis is expected. Therefore, precautions are taken to explain what the stakes are and how to interpret the failures. In other words, we followed a very traditional rhetorical trick of "captatio benevolentiae" – enrolling the benevolence – of the participants.

Fig. 6.6 Photomontage to present the game in the reports – RNRT -Transhumance, 2005

- Second, documents such as manuals or questionnaires actually project the object in a better situation: images can present the objects or its interface in a better light.
- Third, the situation of tests is also organized as a festive and rather congenial event: drinks, food, are provided. Following a recent conference on innovation, I was struck by the obvious and guaranteed consequence of all research projects: a boom in the pizza industry.

Starting a debate with an object would therefore mean at least two things:

- the object foregrounds its unfinished status and therefore invites people to step in and complete it;
- the situation of communication organizes the contributions through the apparatus of research (questionnaires, observers, testers) and a meta-communication on research work (its difficulty, its challenges, its expectations, and the pizzas).

Even though the actors do not use the concept of debate, I think it an accurate way to describe the arguments and counterarguments that are at the basis of tests. I suppose that this word with its political connotations embarrass people in science. In contrast, "critical designers" and some researchers involved in "critical engineering" deliberately qualify debate as their goal. The next section of this chapter therefore compares two "critical trends" within the design and within the engineering communities.

6.4 Critical Design: Designers Questioning Their Contribution

In this section, I describe a trend called "critical design" that started about 20 years ago. In the late '90s in England, at the Royal College of Art, several designers started questioning the place of design in society. They were looking at the tangled web of science, industry, and market economy and started worrying that their contribution was yet another way to feed new products to an already over consuming society. Also, they felt that the speed at which technologies were introduced into the world did not leave time to discuss the pros and cons. A critical posture was called forth that provided with a reflective stance within the discipline itself and not outside of it. These designers were critical of their own mostly unquestioned surrender to scientific "progress". Designers therefore developed critical "products" to influence and change their own discipline from the inside. By studying their claim and their production, I will further reflect on the place of debate within science and design, and how designers develop their own ways to evaluate but also expand the scope of invention.

6.4.1 Design Exploration as a Step Beyond Sociology of Technologies

As pointed out at the beginning of this chapter, designers do not primarily generate debates. Political parties, journalists, lobbyists, and political activists are the usual actors of debates in society as they reach out to a larger audience, organize representations and mediations so that problems and solutions can be discussed and decisions made. But, in this section, I am interested in capturing the special contribution of designers to public discussions and controversies. First, this critical movement has been taking place in a broader trend of design and engineering research that focus on reflexive design and approaches that are good at "problem-setting". Trying to define these design research practices, Fallman later characterized them as "Design exploration" research. "Design exploration" research "often seeks to test ideas and to ask "What if?"—but also to provoke, criticize, and experiment to reveal alternatives to the expected and traditional, to transcend accepted paradigms, to bring matters to a head, and to be proactive and societal in its expression."[56] A sign of recognition of such activities, as Daniel Fallman points out, is "exploring possibilities outside of current paradigms". Design exploration research partly aims at engaging people in reflection, conversation, and debate on the current state of things and on possible alternatives. "In this sense, design exploration is a way to comment on a phenomenon by bringing forth an artefact that often in itself, without overhead explanations, becomes a statement or a contribution to an ongoing societal discussion."[57]

[56] Fallman (2008).

[57] Ibid.

Within the design community, the "critical framework" has been introduced and formalized by Antony Dunne at the Royal College of Art in London in 1999. Anthony Dunne first coined the term in his PhD thesis, "Hertzian Tales: electronic products, aesthetic experience and critical design", later published.[58] The purpose of the book was "to set the scene for relocating the electronic product beyond a culture of relentless innovation for its own sake [...] to a broader context of critical thinking about its aesthetic role in everyday life".[59] Dunne considered that the role of industrial designers was limited to the development of electronic products, while "design, too, has much to contribute as a form of social commentary, stimulating discussion and debate among designers, industry, and the public about the quality of our electronically mediated life".[60] In the 2005 preface, Dunne admitted that "Design is not engaging with the social, cultural, and ethical implications of the technologies it makes so sexy and consumable".[61] The focus of critical design is therefore not so much about what other industrial or civil actors do, as it is to reflect on the role of design at the interface between the industry and society. Dunne and other designers claiming this critical stance are concerned about the fact that design streamlines industrial production with very little consideration for its impact and very little interest in the ethical responsibility of designers. More generally, there is an underlying unease about the hidden agenda that design and science pursue.

As a consequence, design methods and goals are deeply questioned. Among other things, a new figure emerges, very far from the adjusted user that most technology oriented research looks for. Critical designers target the participation of a public who is invited to react – and not only use – these artifacts. They aim at talking to the active reflexive citizen.

Along with the sociology of sciences and technologies (STS), critical designers consider that science is not built in the sanctity of a lab devoid of all social, financial, or political stakes.[62,63] They consider that technology is defined by dynamics involving humans and nonhumans, researchers and artifacts, fictions and tests, values and "the outward clash of reality" (Peirce). However, while they appreciate that STS researchers provide analytical perspective, they also point out that sociologists do not imagine nor build alternatives to the lack of public participation in technological development.

> STS's methods tend to be descriptive, focused primarily on producing accounts of technology and science in action. As a consequence, they have not been applied in any concerted fashion to examining possibilities for the future.

---

[58] Dunne (2008).

[59] Ibid.

[60] Ibid. introduction p. XVI.

[61] Ibid. preface p. XII.

[62] Gaver (2012).

[63] James Auger at RCA, or Alex Taylor at Microsoft research Cambridge for instance see EEAST, 2010, Practicing science and technology, performing the social, Trento, Italy

Critical or speculative designers consider that they can organize public "engagement" and learn about technology from a broader point of view than the "simple" anthropological observation of scientific labs. They want to produce tangible artifacts, that expand the scope of the research, embody the possible consequences, and simulate possible alternative uses. In other words, they create a specific scientific space: the space of sciences design. From an epistemological viewpoint, this activity can be considered as normative and not properly "scientific". However, I think that they also perform a theory of technology: technology is not only "invented" but also "discovered" through debate.

From a design theory perspective, we mostly need to know how they contribute to the invention, concretization, and knowledge of new artifacts in their own way. In other words, we need to look at the production and claims of these designers. Out of multiple projects that can be classified as belonging to the critical trend in design, I chose to study Tobie Kerridge's Biojewelry project, first because it is the product of a collaboration between design researchers, bioengineers and "laypeople", and second because Kerridge also organized a very systematic program of engagement with audiences.[64] Finally, this project is interesting to analyze because, thanks to Kerridge we can access a very rich documentation about the way the biojewelry was made, exhibited, discussed: files, research funding documents, recruitment information, posts, letters, photographs of process and exhibitions, etc.[65]

6.4.2 *Tobie Kerridge – The Biojewelry Project*

Tobie Kerridge is a researcher at Goldsmith College in London and has worked on numerous projects that are designed to engage people in debates about emerging technologies and evolutions in society.[66]

The Biojewelry project has been supported by the National Coordinating Center for Public Engagement in the UK.[67] This is an important institutional fact that gives the project its political background and contributes to its legitimacy. The NCCPE

[64] http://www.biojewellery.com/ (accessed in 2015- no longer accessible)

[65] "Biojewellery is a collaborative project involving Tobie Kerridge and Nikki Stott, design researchers at the Royal College of Art, and Ian Thompson, a bioengineer at Kings College London, its aim is to bring the medical and technical processes of bioengineering out of the lab and into the public arena".

[66] See also Tobie Kerridge's PhD thesis: Designing debate: the Entanglement of Speculative Design and Upstream Engagement, Goldsmith College, July 2015.

[67] http://www.publicengagement.ac.uk/

was started in 2008 after a series of reports and alarms about the gap between higher education, research and the general public.[68] These reports pointed out the economic impact of a lack of collaboration between UK universities and the industry. They also noted that there could be more initiatives of public engagement, if researchers were rewarded when they spent time doing so:

> The Royal Society's report 'Survey of factors affecting science communication by scientists and engineers' in 2006 found that 64% of scientists said that the need to spend more time on research was stopping them getting more engaged and 20% agreed that scientists who engage are less well regarded by other scientists.

The United Kingdom therefore organized a series of actions to support universities that wished to develop public interactive activities, starting with signing a Manifesto for Public Engagement:

> Public engagement describes the myriad of ways in which the activity and benefits of higher education and research can be shared with the public. Engagement is by definition a two-way process, involving interaction and listening, with the goal of generating mutual benefit.

They also listed possible activities, advantages, and partners.

The Biojewelry project benefited from this institutional and political background also because it got funding from the Engineering and Physical Science Council (EPSRC) that was committed to the Public Engagement Program.

The context was therefore very supportive and the project was developed sharing the vocabulary of "engagement". In particular, the NCCPE stressed the fact that engagement worked both ways and therefore expected some benefits for the universities in terms of feedback and shared knowledge. The same argument was taken over on the Biojewelry website:

> This project is at the most contemporary end of a long history of initiatives which aim to create a debate with the public. Successful debates provide a wide audience with detail about the motives, processes and outcomes of science research, and in return the research becomes enriched by the responses, ideas and questions of this audience.

It is precisely this expansive part of the project that I want to look into in this section. I think that the project can be considered on two levels:

- The goal of the project is to engage people with science and therefore it uses communication and mediation tools to exhibit, explain, and explore the subject with different audiences. It is part of a program of popularization and as such

[68] [1] 'Knowledge Exchange between Academics and Business, Public and Third Sectors,' Maria Abreu, Vadim Grinevich, Alan Hughes and Michael Kitson, uk-irc, (PDF)

[2] 'Excellence in Science: Survey of factors affecting science communication by scientists and engineers,' The Royal Society, 2006, (PDF)

[3] 'Report and action plan from the Science for All Expert Group,' BIS, 2010, (PDF)

[4] 'Public Culture as Professional Science: Final report of the ScoPE project (Scientists on public engagement: from communication to deliberation,' Kevin Burchell, Sarah Franklin and Kerry Holden, September 2009, (PDF)

uses rhetorical strategies to simplify, exemplify, and relate biotech to peoples' lives and literacies.[69]
- it is also a research through design project. In that sense, it develops specific means to expand the realm of biotechnologies. In other words, it contributes to the development of the technology by providing original outcomes.

The designers therefore place themselves in a tradition of popularization of science but also consider another aspect and probable specificity of the "collaboration":

> Our backgrounds, interests and previous work provide this collaboration with some unusual features, which we hope will engage an audience in an exciting way.

The whole project is documented on the website and I will only underline some aspects of it. The project started with looking for couples who would agree to give bone cells that the team would make "grow" and turn into rings. The project attracted immediate attention from the press and the New Scientist published an article explaining the process and purpose of the project.[70] It was also noted that they needed an agreement from the UK ethical commission: "The tricky part is that the lucky couple will have to provide bone cell samples, for which the team will get ethical approval only if both people already need surgery. The most likely scenario is that both will need to have a wisdom tooth extracted".

The posts are not only about the biological issue – ethical or otherwise. They are posts on:

- definition of cells and explaining different types of cell cultures
- the design process: the use of 3D modeling, the material of the ring,
- the actual design of the rings (discussed with the couple)
- tools: bioplotter 3D printer[71]
- ethical issues and procedures: "The RCA research ethics guidelines can be downloaded here and the Medical Research Councils guide on tissue samples for research can be downloaded here".
- the recruiting process and how people responded to the proposition which was already part of the debating part of the project: "If this fascinates you, let us know why you would be interested in doing this, and something about the relationship to your partner. If this is not for you, let us know why?" [...]
- "Both of us felt that we'd love to be involved in something that is both so scientifically groundbreaking and that we thought would be so meaningful for the both of us. We're not completely sure of our reasons but we did know that we

[69] Jeanneret (1994).

[70] Jenny Hogan, Cultured bone offers novel wedding rings, New Scientist, 26 February 2005 http://www.newscientist.com/article/mg18524884.900

[71] http://envisiontec.com/products/3d-bioplotter/

wanted to at least register our interest in becoming volunteers since this is a certainly a once-in-a-lifetime opportunity and something we'd love to be part of."[72]

- the circulation of the project information after the two articles in Bizarre Magazine, and the New Scientist. In particular, they follow how different magazines or newspapers pick up the information, one after another, and the consequences in terms of recruitment.
- scientific and artistic dissemination: "conferences and exhibitions May 24, 2005
- The Science Communication Conference: Nikki and I gave a poster describing the project at the BA's 4th Science Communication Conference. The BA is a body tasked with raising public understanding of science work."

6.4.3 Designing for a Design Space "Out of Place"

This particular example is strikingly similar with what we already saw with the art and research examples: an object is designed that occupies a design space that is out of place. Bone cells are being sampled and grown for health reasons and suddenly the same biotechnology is used on the one hand for a rather frivolous purpose: jewelry, and on the other hand for a rather heavy commitment: marriage. Displacing technology from the scientific to the mundane is part of the process that can be qualified as the uncanny,[73] that is both familiar and unfamiliar. The practice seems "cannibalistic", "fetishist", "primitive". And at the same time, it seems really high-tech, scientific, clean. As we have seen in Chap. 5, the design practice here consists in giving attributes coming from two different worlds to an object, that consequently changes the values and perception that we have of biotechnology as well as our representation of more ancient ways of playing with body fragments and samples (in masks for instance). Suddenly we look at biotechnology as if it could be fetishist!

What is remarkable about this experience is that the debating aspect of the project has been thoroughly pursued. The letter to recruit the couples is also an invitation to react to the project. During the exhibitions (for instance "Design and the Elastic Mind", Museum of Modern Art, New York, United States, February – May 2008), the team interviewed people and organized debates. All the information was disseminated in the press.

This is a good example of what Ilpo Koskinen qualifies as design in the showroom[74] that is to say design that concentrates on building an audience rather than users. The characteristics of these designs is to contribute to the expansion of knowl-

[72] http://www.biojewellery.com/project2.html (accessed in 2015 no longer accessible)

[73] Gentes and Mollon (2015).

[74] Koskinen et al., *Design Research Through Practice*.

edge in the sciences by using a strategy of involvement rather than popularization. Critical design explores the potential of technologies by crafting specific artifacts and interactions with these artifacts that do not predict nor prescribe their use but "preview" situations of use with an engaged audience.

6.5 Conclusion: The "Thing" Reopening the "Object"

Despite all the designers' efforts, there is a "criticism of critical design"[75] as being not critical enough or inefficient at creating a real social debate about the emergence of technologies. This criticism also focuses on how, so far, critical practices have been insufficiently exposed so that other members of the HCI community, for example, could take them on and use them as a way to improve their contributions.[76] I contend that the question has been stated in a way that makes it difficult to practice critical design and that the answer lies in a careful appraisal of objects and mediation, a factor which is often neglected. First, the circulation of ideas and the discussions on aesthetic evolution, political dilemma, technical problems, social issues, and scientific foundations are embedded in designs and in mediation. Second, between different exhibitions or tests, designers and researchers may choose different versions of their things and different mediating strategies. Things do not have a fixed identity because, in this perspective, they are, "re-debated" in space and time. The expression "design as debate" therefore applies to:

- Debate within the artifact itself: things have a way of creating some controversies in particular when they do not belong to any clear-cut category that makes them easy to qualify. This is definitely the case of technical invention or critical design.
- Debate amongst social actors: a thing is presented in different situations that mediates the way it is introduced in society. Each occurrence is the occasion for a renewed discussion about its values.

In any event, the thing is presented as an unknown, something that is lacking in depth, precision, purpose. The "openness" of the artifact – that calls for a contribution – is always context dependent and can be staged in different ways. In the art example, it was an interesting organization of quotation and *ekphrasis* (that is the description of a visual artifact with words). In research, the object is presented as lacking completeness and just a stage in the research process. In critical design, the

[75] Bardzell and Bardzell (2013).

[76] Bardzell et al. (2013).

"Thing Staging"	art	Openness by remediation, quotation, ekphrasis, commentaries
	research	Openness by introduction in the ongoing research process
	Critical design	Openness by circulation and restaging

Fig. 6.7 Thing staging: overview of openness strategies

object is exhibited and mediated in a number of places that foreground its questionable meaning (Fig. 6.7).

In the next sections of this conclusion, I review some aspects of these strategies as well as elaborate on the debating aspects of "critical technical practices". In my opinion, these debates supported by design are not only important for a socially responsible practice, they also contribute to the expansion of the invention and correlated knowledge, in other words they are truly generative. I suggest that it is possible because design/practice now not only considers "objects", it turns to making "things".

6.5.1 Critical Strategies and the Incomplete Thing

In all these examples, we clearly see different "internal narratives" and rhetorical strategies that converge to expand the meaning of artifacts. Internal narrative strategies trigger the audience's attention and subjective concerns. Alex Seago and Anthony Dunne speak of the object as discourse.[77] I call "internal narrative" the way formal aspects stage a theme and a form of reception: for instance, a system localization that is not entirely working because a faulty interface. In the engineering examples, this internal narrative is a cue for the reflexive experience of the testers that they need to think about mobility, sharing information, and hybrid man/machine infrastructure. But these internal narratives are not enough to support a discussion. I call rhetorical strategy the way these things are introduced to their audience. Rhetorical strategies encourage people to interpret and contribute to a new definition of what they see. In the art example "Tell me your secrets", the internal narrative is about intimacy and the rhetorical strategy consists in framing or "un-framing" the obscene in the art world thereby giving it a different meaning. In the research projects, demonstrators are deliberately incomplete. They are supported by a rhetorical strategy that legitimizes this incompleteness by a discourse on science and the organization of testers' contributions. In critical design, the main narrative is based on the "uncanny" which is both familiar and unfamiliar. People recognize the objects but they discover that they are twisted and that they uncover uncomfortable

[77] Seago and Dunne (1999).

	Object and main internal narrative	Mediation and main rhetorical strategy
Art example	Obscene	Support / organize the art world
Research example	Unfinished	Engage the audience to contribute and craft a culture of research
Critical design example	Uncanny	Unsettle and reassure the audience

Fig. 6.8 Overview of internal narratives and rhetorical strategies

associations of ideas. The rhetorical strategy consists of giving people a chance to debate this feeling of uncanniness and discomfort (Fig. 6.8).

The difference between debate in general and debate with design is that design or art or engineering research offer a chance to actually manipulate the object, to get a feel for it and not only to work on its representations. Critical designers insist that their debates have to be based on an artifact. For instance, Dunne states that "our ideas make their way into the material world in some way; it's not enough that they end up as pure thoughts. They must be embodied in object typologies that we understand: furniture, products, clothing, buildings…"[78] Objects matter because they are not a simple transcription of a technology but on the contrary a quest to expand its potentials. The role of the experimentations is to give agency to objects and to surprise people in how they can be affected. The debates are so to speak tangible and embodied. However, as we have seen, these properties do not close the definition.

In the different examples of this chapter, it is obvious that designers and researchers want to make up for a deficit of intelligibility of emerging technologies. Designer and theorist, Augusto Morello[79] (but also Manzini,[80] Hatchuel et alii[81]) describe the design challenge of our contemporary society as a general identity crisis. The response to such a situation is not only formal, as one might think, but communicational as well. Critical design shows that no object stands alone and that, in the case of radical innovation, design practice includes – implicitly or explicitly – the design of the presentation of these things. This was done in the case of "Tell me your secrets" because of the censorship. This is also done in research projects.

6.5.2 "Critical Technical Practice": For a General Theory of Scientific Change

All critical strategies are both destabilizing and open to reflection in different ways. In this respect, critical design is part of what Philipe Agre called "critical technical practices", that is practices that help define new episteme. I want to suggest that his

[78] Rickenberg (2008)

[79] Morello (2000).

[80] Manzini (2009).

[81] Hatchuel et al. (2014).

contribution is not only helpful because it qualifies certain design and engineering practices. It is also helpful because he gives a framework for understanding how critical practices supporting design as debate are necessary to innovation.

In the '80s, in engineering sciences, critical engineering was introduced and discussed by Philipe Agre as the "Critical Technical approach" and was further pursued by scientists in information sciences, sociology of technologies, and in the field of HCI like Paul Dourish,[82] or Höök, Sengers and Andersson doing work on the "influencing machine".[83] They questioned the social worth of their production:

> An alternate approach for supporting more authentic interaction with and around technology does not attempt to fit complex experiences into computer models but, rather, looks at ways in which technology can stimulate reflection on, enhance awareness of, and create opportunities for meaning making activities.[84]

The question was not only raised in terms of ethics. The authors contended that their own discipline could be blind to its own limits, therefore ignoring possible innovative approaches.

Agre's text – that presents autobiographical elements about his career in Artificial Intelligence – analyzes what he sees as the limitations of AI as a discipline. While he shows that it has a generative capacity for opening new fields and models, he also shows that it is lacking in several respects. First, he argues that AI has no space for self-reflection on its methods and concepts. This is detrimental to the discipline because it limits its capacity to evaluate its findings. It also limits ethical discussions about the systems that it produces. Finally, it restricts the dialog that AI could have with other disciplines and that could help expand its own understanding and scope. Second, he points out characteristics of AI that hover between science and engineering.[85,86] and tend to avoid questions about the validity of its concepts (supposedly based on an understanding of phenomena) by providing a technical answer: my system works better than yours. "AI projects are sometimes scientific in intention, sometimes engineering, and sometimes they shift subliminally from one to the other". While this assessment of the discipline may no longer apply, we find that Agre's criticism exposes a dilemma in engineering research. On the one hand, the metaphorical nature of concepts is precisely what helps the discipline expands its goals and productions. On the other hand, it does so without the benefit of assessing the metaphorical process at play. This means that the relevance, limits, but also benefits of the metaphors cannot be fully appreciated nor explored.

Finally, Agre considers that the field's use of language can make it extremely difficult to come up with radical innovations that can be considered as such. "AI's

[82] Dourish (2004).

[83] Höök et al. ii, (2003).

[84] Boehner, critical technical practice.

[85] Armytage (1966).

[86] Auyang (2006).

elastic use of language ensures that nothing will seem genuinely new, even if it actually is".[87] He also points out the fact that AI frames the way questions are asked in terms such as modularity, learning, hierarchical plans: "AI's intricate and largely unconscious cultural system ensures that all innovations, no matter how radical the intentions that motivated them, will turn out to be enmeshed with traditional assumptions and practices." In his narrative of self-discovery and change, Agre suggests three reasons to explain why this framework finally failed him: first, the impression, as a young researcher, that he could never do something new; second, the contrast between the models of human phenomena in AI and his own documented experience of everyday life; and third, the reading of a number of philosophers and social scientists, in particular Foucault. This last fact is interesting because it appears that Foucault's structuralism presented enough similar formalism with AI that it helped bridge the gap between social science and Agre's own scientific background. The benefit of a critical position is therefore, according to Agre, primarily related to the capacity for innovation and, secondly, evaluation. A critical position is a position that questions the meaning of schemata and concepts so that it reopens the possibility of new definitions and new models to emerge. A critical position also looks into the metaphorical process so that "dead metaphors" can be discarded to give way to new metaphors. Easier said than done according to Agre, who is suspicious that creating new metaphors might never happen because of the weight of tradition and intellectual format. He therefore suggests that hermeneutics could be a method. His interest in hermeneutics is based on interpretation that he wants to apply to things that do not work. He suggests that we should look at technical difficulties as signs of deeper problems rather than superficial glitches:

> Perhaps we can learn to approach technical work in the spirit of reductio ad absurdum: faced with a technical difficulty, perhaps we can learn to diagnose it as deeply as possible. Some difficulties, of course, will be superficial and transient. But others can serve as symptoms of deep and systematic confusions in the field.[88]

In the hermeneutic framework, interpreting technical problems means that they are considered as a web of interrelated nodes that build a hidden structure that the analysis uncovers.

> Research could proceed in a cycle, with each impasse leading to critical insight, reformulation of underlying ideas and methods, fresh starts, and more instructive impasses.[89]

Agre therefore advocates that the critical technical approach could involve disciplines that usually look at meaning-making such as humanities and social sciences. This has obviously been the case for this book which tries to look at design (and conception more generally) through humanities. More generally speaking, the question is who is invited to debate so that new episteme can emerge?

---

[87] Agre (1997).

[88] Agre (1997).

[89] Agre (1997).

6.5.3 *Ethical and Political Stakes: Who Can Debate?*

Through the evolution of science and technology, every step can impact in ways that cannot be fathomed by single actors. The kind of issues that we face today (pollution, radioactivity, financial crises, unemployment, etc.) are therefore beyond any individual decision and need to be discussed globally. The complexity is not only due to the scale of problems but is also related to the fact that no discipline can claim that it can encompass the whole issue. The concept of multidisciplinarity might not even be sufficient because the addition of knowledge does not necessary allow for a synthetic view of global issues. Traditional epistemologies of sciences are therefore compelled to reconsider the way they look at their interdependency and how they have to admit to knowledge and non-knowledge. I will develop these ideas in Chap. 7.

In A Society of Risk?[90] the question is largely about where and who can discuss what sciences are about. The question was previously asked by philosophers like Lyotard who studied the binds between political powers and scientific activities. In 1979, Lyotard[91] noted that, in the age of the computer, the relations between the world of science and the world of politics were more than ever conflicting about the definition of what we "must" know and who is empowered to decide. Information technologies reopen the need for a definition of what knowledge is and who has it. While the numerous discourses about collective intelligence, open innovation, are cases in point, the complexity and interrelatedness of all the projects make it difficult to organize a debate about what research findings are worth for society. In his analysis of the Public Sphere and of the relation between science and society, the German philosopher, Jürgen Habermas alerts us that it is critical that we invent institutions where a debate about the value and the use of science can take place. In my opinion, critical design and critical technical practices are spaces that allow for such discussions. This explains why critical designers repeatedly reaffirm that they are designers and that their work is to be evaluated within design parameters and not artistic parameters even though their productions have been compared. I think that their claim makes sense because artists and designers do not question the audience in the same way and they present their work in different institutions. Designers insist that they are not artists because they want to use the "institution" of design to question design, science, and industry, and their relation to society at large. They want people to play the game of a "ready to market" product, so that they consider how they are going to be personally affected by the development of certain technologies. Designers therefore have become activists who challenge the applications of contemporary sciences and also help to change their paradigms. Critical design and critical engineering practices are responding in their own terms to the way science moves on and affects the world.

---

[90] Beck (1992).

[91] Lyotard (1979).

6.5.4 *From Objects to Things: The New Design Rationale*

Design as debate relies on a theoretical model of objects. Within this paradigm, there is no ontology of objects, the object is not defined by a set group of fixed properties, and therefore, there is no technological determinism. Within design as debate, the object is also not the product of a purely subjective experience impossible to share with other people. The design as debate paradigm recognizes formal qualities and experiential properties. It puts in the foreground a theory of "the thing" that is at the center of collective discussions and experiences. I have been surprised to see that in a lot of the design research, statistics are considered the best way to insure some kind of shared collective subjectivity around the experience. But that means that the expansive properties of debates are forfeited in the name of one good average solution for all. The different examples either in art, design, or engineering research show that debates are actually useful in that they are collective **and** expansive. "Debates" are a defining process relying on rational arguments: an argumentative public sphere. In this public sphere, each actor presents his/her rational arguments to defend a point of view, a definition of the thing. These discussions may include arguments about shapes, formats, aesthetics, but also function, purpose, usefulness, as well as ethics, responsibility, sustainability. The arguments presuppose what Habermas[92] describes as a "horizon of mutual understanding". The participants have not only a right but a capacity to argue rationally to get to a common definition, a capacity that is at the basis of their everyday organization (Habermas uses the findings of ethnomethodology). In the field of art, Rochlitz[93] (opposed in this respect to other theoreticians of art such as Genette,[94] or Schaeffer[95] who consider aesthetic as a personal relationship to art) offers to consider art as the space for such a debate. Rochlitz does not deny the fact that interests (personal, institutional, economical) are involved in these debating rings, but he contends that this does not prevent the building of an argumentative space that is different from the sheer imposition of ideas by threat or force.

Debates that question the identity of artifacts are then central for designers who now want to design "things" before designing "objects". Things and objects are very different concepts that redefine what design can be about. Comparing their etymology and their use in the context of their invention helps us consider how we gradually delineated design.

---

[92] Habermas (1985).

[93] Rochlitz (1998).

[94] Genette et Goshgarian (1997).

[95] Schaeffer (2000).

The word "object" is based on the Latin root of the verb "to throw" and the preposition "ob": "in front of". The object is thrown in front our eyes and senses. In the dictionary of etymology, we learn that the late fourteenth century word means: "tangible thing, something perceived or presented to the senses". It comes from the Medieval Latin "objectum "thing put before"(the mind or sight), and from the verb obicere"to present, oppose, cast in the way of". Because it is thrown in our path, in front of us, we pay attention to it. Gradually, the uses of the word converge on the tangible characteristics as they are presented to the senses. It is therefore a word with two sides: on the one hand, it points to the fully autonomous material obstacle. On the other hand, it points out to the experience of the beholder. The object is also primarily conceived as the subject of scrutiny, of exploration. It presupposes that we can "know" it because of the distance and the play on our senses. This definition is helpful as it insists on the materiality and on our capacity to experience the object through a variety of senses. An object is eventually also an object of knowledge: its identity is fixed and its properties can be described by a discipline.

The word "thing" is quite different. The Old English þing first means "meeting, assembly," later "entity, being, matter" (subject of deliberation in an assembly). If a meteorite falls in the middle of a village, people gather around it and together try to figure out what this "thing" is. The thing is therefore impossible to name. Its identity is questioned and to name it, the community needs to convene and to argue about it. What is more, the word "thing" is not limited to tangible artifacts. It can apply to ideas, events, organizations. In fact, it focuses on the social aspects of our relation to the world on at least two levels. Things that we cannot name yet with precision encompass all that can affect us. So, the discussions are shared because the whole community can be affected. Second, things are socially defined. They are what we discuss together to make sense of it. A thing is of interest to a community that discusses it. Thing is therefore a political word. The French word "chose" from the Latin "causa", goes a little be further in the same direction. A "causa" is a "judicial process, lawsuit, case". The community has to pass a judgment about the thing, after listening to various opinions, including demonstrations but also ethical concerns. In any event, the task of naming things or deciding on what to do with them or how to judge them has to be decided by a community not by the clear-sighted observation of a sole observer. Used colloquially since c.1600 to indicate things the speaker cannot name at the moment, often with various meaningless suffixes, e.g. thingumbob (1751), thingamajig (1824). It does not originally mean that there is a lack of observation but rather that there is a lack of communication and mutual understanding.

When design/conception switches from objects to things, it has to rethink the materiality and purpose of the artifact. It does not only describe and put together the object, it questions the thing to be. This definition of design/conception signifies that the making of an X, an unknown, goes with a destabilization of bodies of knowledge that supported the existing objects. To do so, design uses multiple disciplines. In the next chapter, I want to describe how design is fundamentally in-disciplined because it disrupts the existing disciplines.

References

Agre, P. E. (1997). Toward a critical technical practice: Lessons learned in trying to reform AI. In G. Bowker, L. Gasser, L. Star & B. Turner, (Eds.), *Bridging the great divide: Social science, technical systems, and cooperative work*. Erlbaum.
Armytage, W. H. G. (1966). *A social history of engineering* (2nd ed.). Cambridge: The MIT press.
Auyang, S. Y. (2006). *Engineering: An endless Frontier*. Cambridge: Harvard University Press.
Baldwin, C., Hienerth, C., & Von Hippel, E. (2006). How user innovations become commercial products: A theoretical investigation and case study. *Research Policy, 35*(9), 1291–1313.
Bardzell, J., Bardzell, S. (2013). What is critical about critical design? In *Proceedings of the SIGCHI Conference on Human Factors in Computing Systems* (pp. 3297–3306).
Bardzell, S. et al. (2013). Critical design and critical theory: The challenge of designing for provocation. In *Proceedings of the Designing Interactive Systems Conference* (pp. 288-297).
Battarbee, K., & Koskinen, I. (2005, mars). Co-experience: User experience as interaction. *CoDesign, 1*(1), 5–18.
Beck, U. (1992). *Risk society: Towards a new modernity* (1st ed.). London: Sage.
Becker, H. S. (1982). *Art worlds*. Berkley: University of California Press.
Boehner, K., et al. (2005). Critical technical practice as a methodology for values in design. *CHI 2005 Workshop on quality, values, and choices*. Portland, OR, April 2–7.
Coleridge, S. T. (1985). *Biographia Literaria: The collected works of Samuel Taylor Coleridge, biographical sketches of my literary life & opinions* (First Paperback Edition). London: Princeton University Press
Cross, N. (Ed.). (1971). *Design participation: Proceedings of the Design Research Society's conference*. Manchester: Academy Editions.
Dourish, P. (2004). *Where the action is: The foundations of embodied interaction* (New Ed ed.). Cambridge: The MIT Press.
Dunne, A. (2008). *Hertzian Tales: Electronic products, aesthetic experience, and critical design*. Cambridge, The MIT Press.
Eckert, C., & Stacey, M. (2000). Sources of inspiration: A language of design. *Design Studies, 21*, 528.
Fallman, D. (2008). The interaction design research triangle of design practice, design studies, and design exploration. *Design Issues, 24*(3), 4–18.
Gaver, W. (2012). What should we expect from research through design? In *Proceedings of the SIGCHI Conference on Human Factors in Computing Systems*. 937–946. http://dl.acm.org/citation.cfm?id=2208538.
Gaver, W., & Dunne, A. (1999). Projected realities: Conceptual design for cultural effect. In *Proceedings of the SIGCHI conference on Human factors in computing systems* (pp. 600–607). http://dl.acm.org/citation.cfm?id=303168
Gaver, W. W., et al. (2004). The drift table: Designing for ludic engagement. In *CHI'04 extended abstracts on Human factors in computing systems* (pp. 885–900). http://dl.acm.org/citation.cfm?id=985947
Gaver, W., et al. (2010). The prayer companion: Openness and specificity, materiality and spirituality. In *Proceedings of the SIGCHI conference on Human factors in computing systems* (pp. 2055–2064). http://dl.acm.org/citation.cfm?id=1753640.
Genette, G., & Goshgarian, G. M. (1997). *The work of art*. Ithaca: Cornell University Press.
Gentes, A. (2001). Les sites artistiques sur Internet : essai d'une typologie des écrans d'accès aux oeuvres d'art. *Solaris, Matière numérique: la production et l'invention des formes, 7*. http://gabriel.gallezot.free.fr/Solaris/d07/7gentes.html (retrieved 20 September 2017).
Gentes, A. (2003). "Art-titres" sur Internet : enjeux de l'énonciation curatoriale. *Communications et Langages, 137*, 88–100.
Gentes, A. (2007). L'intime à l'épreuve du réseau. *Communication et langages, 152*, 89–105.
Gentes, A., & Mollon, M. (2015). Critical design: A delicate balance between the thrill of the uncanny and the interrogation of the unknown. In D. Bihanic (Ed.), *Empowering users through*

design: Interdisciplinary studies and combined approaches for technological products and services (pp. 79–101). New York: Springer.
Grenier, C. (2008). *La revanche des émotions: essai sur l'art contemporain*. Paris: Seuil.
Habermas, J. (1985). *The theory of communicative action, Volume 1: Reason and the rationalization of society* (trans: McCarthy, T.). Boston: Beacon Press.
Hatchuel, A., Weil, B., & Collectif. (2014). *Les nouveaux régimes de la conception : Langages, théories, métiers*. Paris: Editions Hermann.
Höök, K., Sengers, P. et Andersson, G. (2003). Sense and sensibility: Evaluation and interactive art. In *Proceedings of the SIGCHI conference on Human factors in computing systems* (pp. 241–248).
Hymes, D. H. (1974). *Foundations in sociolinguistics; an ethnographic approach*. Philadelphia: University of Pennsylvania Press.
Jeanneret, Y. (1994). *Ecrire la science – Formes et enjeux de la vulgarisation*. Paris: PUF.
Koskinen, I. K., et al. (2011). *Design research through practice: From the lab, field, and showroom*. Waltham: Morgan Kaufmann Publishers Inc.
Le Masson, P., Weil, B., & Hatchuel, A. (2017). *Design theory. Methods and organization for innovation*. Cham: Springer.
Luck, R. (2003). Dialogue in participatory design. *Design Studies 24*(6, novembre), 523–535. doi:https://doi.org/10.1016/S0142-694X(03)00040-1.
Lyotard, J.-F. (1979). *La condition postmoderne* (Editions de Minuit ed.). Paris, Les Éditions de Minuit.
Lyotard, J.-F. (1984). *The postmodern condition: A report on knowledge* (1st ed.). Minneapolis: University of Minnesota Press.
Manzini, E. (2009). New design knowledge. *Design Studies, 30*(1), 4–12.
Morello, A. (2000, Autumn). Design predicts the future when it anticipates experience. *Design Issues, 16*(3), 35–44.
Newton, S. (2004). Designing as disclosure. *Design Studies, 25*, 93–109.
Prahalad, C.K., & Ramaswamy, V. (2004). Co-creation experiences: The next practice in value creation. *Journal of Interactive Marketing, 18*(3, Janvier), 5–14. doi:https://doi.org/10.1002/dir.20015.
Rickenberg, R. (2008, Spring). Interpretation, collaboration, and critique: Interview with Anthony Dunne. *Journal of Design Management, 3*(1), 22–28.
Rochlitz, R. (1998). *L'art au banc d'essai: esthétique et critique*. Paris: Gallimard.
Sanoff, H. (2000). *Community participation methods in design and planning*. New York: Wiley.
Saper, C. J. (2001). *Networked art*. Minneapolis: University. of Minnessota Press.
Schaeffer, J.-M. (2000). *Adieu à l'esthétique*. Paris: Presses universitaires de France.
Seago, A., & Dunne, A. (1999). New methodologies in art and design research: The object as discourse. *Design Issues, 15*(2), 11. https://doi.org/10.2307/1511838.
Simakova, E. (2010, August). RFID 'Theatre of the proof': Product launch and technology demonstration as corporate practices. *Social Studies of Science, 40*(4), 549–576. first published on 15 June , 2010.
Simondon, G. (2016). *On the mode of existence of technical objects* (trans: Malaspina, C.). Minneapolis: Univocal Publishing.
Smith, W. (2009, June). Theatre of use: A frame analysis of information technology demonstrations *Social Studies of Science, 39*, 449–480.
Stallabrass, J. (2003). *Internet art. The online clash of culture and commerce*. London: Tate Publishing.
Star, S. L., & Griesemer, J. R. (1989). Institutional ecology and boundary objects: Amateurs and professionals in Berkley's Museum of Vertebrate Zoology, 1907-39. *Social Studies of Science, 19*, 387–420.
Von Hippel, E. (2006). *Democratizing innovation*. Cambridge: The MIT Press.
Zwick, D., Bonsu, S. K., & Darmody, A. (2008). Putting consumers to WorkCo-creationand new marketing govern-mentality. *Journal of Consumer Culture, 8*(2), 163–196.

Chapter 7
Conclusion: The Indiscipline of Design

7.1 Introduction: Far Too Many Disciplines or Not Enough?

Most designers claim that their discipline is multidisciplinary and transversal, and that this movement from one discipline to another is what defines design. The literature on design confirms that to understand design per se is sometimes difficult as it seems that it always borrows from multiple disciplines: physics, mechanics, but also biology, chemistry, and numerous fields in life science, as well as social sciences like ethnography or sociology. In both ideation and production processes, design is described as interacting with multiple disciplines (Rodgers and Bremner 2013[1,2] Cross 2001[3]; Dykes et al. 2009[4]; Findeli et al. 2008[5]; Harfield 2008[6]; MacKay and Fayard 1997[7]; McKay and Marshall 2007[8]). This claim takes place both at the institutional level to demand a specific space for a multidisciplinary practice[9] and at the epistemological level to establish the rationale of a discipline while it borrows and modifies concepts and methods from other disciplines as well as develops its own. Elaborating on this research, I also want to take this claim seriously and test the hypothesis that this "indiscipline" of design is not only valuable for the practice of designers but is precisely the rationale of design as the discipline of conception (design/conception).

[1] Rodgers and Bremner (2013).
[2] Bremner and Rodgers (2013).
[3] Cross (2001).
[4] Dykes et al. (2009).
[5] Findeli et al. (2008).
[6] Harfield S. (2008).
[7] Mackay and Fayard (1997).
[8] McKay and Marshall (2001).
[9] Dubreuil (2007).

A. Gentes, *The In-Discipline of Design*, Design Research Foundations,
DOI 10.1007/978-3-319-65984-8_7

There are two aspects to this discussion. The first aspect is to recognize the assembly of disciplines and to understand how this assembly can be generative and expansive. The sociology of science has inspired my approach to observing "how science is made",[10] but first, it does not focus on the aesthetic properties of the actual invention of an X, and second, it does not describe the turmoil of disciplines that seem systematically correlated with design. Classical epistemology is also rather silent on the subject of design and engineering. For Antonie Meijers[11], this lack of recognition is obvious in philosophical publications that rarely provide entries on either technical or engineering philosophies. Anne-Françoise Schmid confirms, through an historical analysis, the lack of reflection on engineering sciences within epistemology.[12] This book does not claim to repair this lack of concern. However, I want to describe how design challenges the status quo of bodies of knowledge thereby allowing for new concepts and artifacts to emerge. To explore this hypothesis, first I use concepts that the philosopher of science, Anne-Françoise Schmid, introduced to understand contemporary sciences and conception: primarily her definition of "under-determination" and her concept of "integrative object". Second, I pick up some examples already studied in the course of this book and in particular the example of the game in the Museum of Arts and Techniques (ANR-PLUG), but I look at them from a fresh standpoint: behind the artifacts, names, stories, I consider what is at stake for the discipline of design. On that basis, I suggest a typology of pluridisciplinarities in design: additive, hybrid, and under-determining. For each type of operation between disciplines I want to show how generative the process is.

The second part of this discussion deals with the humanities. As the previous chapters of this book have shown, the humanities play a fundamental role in the multidisciplinarity of design because they under-determine and are under-determined by other disciplines. Even more profoundly perhaps, the humanities have helped researchers understand the episteme of design as an aesthetic of conception. The concluding remarks introduce the discussion on how to name this expansive multidisciplinarity. I finally chose to call it the "in-discipline" of design after Rancière, but a dialog with Duchamp and Oulipo suggested by the specialist of prospectivism, George Amar, points other interesting options.

7.2 Addition of Disciplines: The "Leonardesque Aspiration"

Alain Findeli observes in his article on theories of design[13] that design is often considered as an applied science and that the number of sciences that can be "applied" by design is almost infinite. Looking at the curricula of the Chicago and Ulm schools of design, he shows that they included more and more disciplines either from hard

[10] Latour (2005).

[11] Meijers et al. (2009).

[12] Schmid (1998a, b).

[13] Findeli (2006).

sciences or social sciences. However, there is no explanation of how design should "logically" be an application of any of these sciences. Qualifying design as an applied science, he argues, is not focusing on design as a discipline with its own epistemology. In addition to what Campbell, quoted by Stein,[14] calls the "Leonardesque aspiration", design is not only given the impossible task to "know it all" but also denied its own episteme. However, design projects do involve multiple disciplines to produce their objects. I want to rapidly see how it all starts either from the object or from the activity.

7.2.1 *What Comes First? Object or Activity?*

In his book on Designing Engineers, Bucciarelli asks his famous rhetorical question: how to define a telephone? This complex device depends as much on the result of signal processing, as that of social uses, antennas, software, regulation of frequency allocations, intense technical standardization negotiations, and highly-virulent political and economic discussions.[15] He then goes on to prove that though the object works, there is not one discipline involved but many and none that can offer a synthetic view of the object. If we want to understand technological pluridisciplinarity, we can therefore start from an object, and count how many disciplines are involved.

In most projects where we wanted to turn places into pervasive environments, we had at least four main disciplines: computer sciences, game design, information and communication sciences, networks. In fact, projects could include up to ten disciplines like ergonomics, sociology, geography, signal processing, electronics, and physics of communication. However, the articulation of these disciplines was not left to chance. In all these projects, the disciplines are organized around the object, by layers that go basically from the hard components of the systems to the application, service, and user. The theoretical architecture of the system therefore holds together the pyramid of disciplines. As long as the "user" is considered as a "layer" of the system, the logic of the architecture is not compromised. Mostly this last "layer" is related to the proper functions of the device, hence to the activity of the user. Of course, one wants to make sure that the different components of the architecture work together. Thus, many models of co-design have been refined that test the functions of two different levels together. But the combination of disciplines is less straightforward than it seems, in particular because some issues are transversal and cannot be solved by one single discipline attached to one single layer or two. More broadly speaking, the ISO (International Organization for Standardization) model of OSI (Open Systems Interconnection) is contested in many instances because it no longer reflects the way systems operate and prevents innovation.

[14] Stein (2007).

[15] Bucciarelli (1996).

However, the OSI of ISO is a convenient even if slightly outdated boundary object that still organizes the way projects are structured and defended.

Another option is pluri-disciplinarity by activity. Many suggest that pluridisciplinary research is done by iteration and somehow trial/error. Wendy MacKay and Anne-Laure Fayard observe that this pluri-disciplinarity is not any kind of bricolage but has to be consistent with the tenants of the original disciplines:

> we must conduct our work in a way that is fundamentally sound at the level of each discipline we draw from and viewed as legitimate by our academic colleagues.[16]

This question is addressed in HCI or information sciences when researchers make observations of original situations, embed hypotheses in artifacts, test these artifacts in the field and then observe the characteristics of the new situation. An iterative process allows the different disciplines coming from engineering sciences and social sciences to adjust to each other, to their findings, productions, and methods. In my experience, the interplay of discipline in that case is ruled by what Isabelle Demeure and I called the "artistic director". In other words, while the project is pluri-disciplinary, the publications are carefully crafted so that they fit with the discipline of the researcher who is temporarily in charge. In other words, every researcher may publish with the help of other scientists but with the agreement that the others will fit his/her lead, for that particular publication. However, in such a model, disciplines are not actually challenged since every researcher can in some way go back to her roots without modifying their fundamental tenets.

We can therefore look at multidisciplinarity by starting from the object and adding disciplinary bodies of knowledge as the project grows. Disciplines are added and organized thanks to an architecture of layers with overlaps (or "triangulation" to use Wendy McKay's expression). Or we can look at the organizational dimension of pluri-disciplinarity and concentrate on how disciplines survive their cohabitation while developing a complex and multidimensional object. In a truly innovative project, I contend that this interaction of disciplines is never without impact on the initial concepts.

7.2.2 *Hybridization of Concepts*

In a pluridisciplinary project, the interplay of disciplines does not leave them, their concepts, and their methods, unchanged. Disciplines influence each other. We need another metaphor to describe this effect: hybridization. I have tackled this issue mostly in Chap. 5, as I looked at the metaphoric process at play in the invention. The oxymoron blends together two concepts to create a new one, but at the same time transforms the original concepts. This process supports the creation of new artifacts but also affects disciplines, their concepts and methods. In an article written with

[16] Mackay & Fayard, HCI, Natural Science and Design: A Framework for Triangulation Across Disciplines.

Fig. 7.1 RIAM-PLUG - Pluridisciplinarity: the humanities, mobile technologies, and new media

Camille Jutant[17] for the journal "Culture and Musee", we described how the concept of "media" acted as the tipping point in a museum project and how every discipline was influenced by the others through the channel of this concept. In particular, computer scientists who first focused on data and information, gradually embedded a larger vision related to their device as a media. They redefined communication as shaped by the technical substrate as well as by the semiotic signs and finally put into practice MacLuhan's principle: "the medium is the message". The different stages of the signs (from their coding by the computer scientists to their display to users) were therefore perceived as different symbolical experiences (Fig. 7.1).

To look at design as a multidisciplinary activity, therefore means to look at how a process of mutual influences and hybridization happens and to analyze the operations that take place to trigger and support this hybridization.

[17] Gentes and Jutant (2012).

However, I want to go a step further. While the discussions about inter-trans or multi-disciplinarity tend to describe the modes of knowledge circulation, it is important to observe how these movements give rise to a new "thing".[18] I want to argue that situations of invention are based on the destabilization of disciplines by removing some of their tenets. In other words, invention is made possible because some design space opens up. Design epistemology is related to this operation of under-determination of a discipline by another.

7.3 « Under-Determination » as the Design Episteme

This section will start with another way to look at the project PLUG and will tell the story of a multidisciplinary scientific collaboration that could have gone awry due to its "undecidable" object. Eric Gressier-Soudan, the CNAM CEDRIC lab manager, wanted to set up a research project within the public funded RIAM[19] program, which, as opposed to the previous programs (RNRT,[20] RNTL[21]) I had participated in, aimed to create a synergy between technologies and cultural productions, as is the case for video games. In other words, this particular research program offered a space for the interaction of disciplines. The technical dimension, as shown in the 2007–2010 report of the national research agency that establishes an overview of its activities, remained nevertheless strong:

> almost 89% of the projects proposed major technological breakthroughs or innovations adhering to the latest global developments. This technological innovation dynamic, that supports the production and distribution of content has, very logically, given birth, directly or indirectly to true innovations in usage.

Based on my previous experiences, I assumed that this new project would support a technical invention. But, at the outset of the project, Eric Gressier-Soudan refused to provide any technical research question. He deliberately kept the terms of the project vague:

> PLUG studies mobile and embedded technologies to implement atmospheric/pervasive/ubiquitous games and their acceptability within a socio-cultural, economical and industrial context. The conception and game-design of pervasive games are the major aspects of PLUG.[22]

The text, which I would finally adhere to, seemed to deliberately confuse the issue: "PLUG intends to use all the technologies available to enable the construction

---

[18] Bremner and Rodgers (2013); Cross (2006); Dykes et al. (2009); Findeli et al. (2008); Gentes, Valentin, Brulé, (2015); Stein (2007).

[19] Recherche et Innovation en Audiovisuel et Multimédia (Audiovisual and Multimedia Research and Development).

[20] Réseau National de Recherche en Télécommunications.

[21] Réseau National en Technologies Logicielles.

[22] RIAM PLUG – document for the call.

of ubiquitous applications". I was skeptical: how could one possibly claim serious scientific research in Information and Communication systems while using "all technologies"? How could a computer lab not have a middleware to explore, an architecture to elaborate or to design a new security system? How could a goal be described as "atmospheric"!? I perceived this lack of proper technical goal as a ruse to enter into a bid for a "content-oriented" project, and expected that at some point or another the supervisor would reveal his technical ambition. He never did! Some technical contours seemed to appear, tracing an unidentified zone to explore, a sort of outline without a core. The usual distribution of disciplines that I presented on different occasions in this book was no longer supporting the project. There was no hierarchy but an assembly of disciplines that would hopefully find a way to adjust to one another at some point to deliver a yet unknown object.

Eventually, two miracles happened. The first miracle was that the project was accepted and got funded. The second miracle was that the team did come up with a truly innovative device and new knowledge that challenged and redistributed the disciplines involved. Thanks to Eric Gressier-Soudan's tenacity, we were to experience a type of invention, under the form of an unknown entity, an "X", that was not even determined by a technical program. While this experience was somewhat special, it made me look back at my other projects and realize that this indeterminacy in the definition of the final goal, as well as the role of design/practices as a dynamic that challenges set disciplines, was at the core of the rationale of design/conception.

7.3.1 "Under-Determination" of Disciplines and "Integrative Thing"

Before looking at PLUG more closely to see how multidisciplinarity can be generative, I want to elaborate on Anne-Françoise Schmid's concept of under-determination. Her contribution is central to this conclusive chapter on design because she focuses on how a new "object" can happen in or in between the scientific disciplines. In other words, she looks at the conditions of emergence of new concepts. Like other philosophers who consider how sciences can evolve,[23] Anne-Françoise Schmid looks at how disciplines can actually move away from their paradigms. If we follow her hypothesis, we understand that design – understood as radical innovation – is central to sciences. This definition is quite different from the addition process usually understood as being the principle of multidisciplinarity. It is different because it focuses first on design, i.e. conception, and second on disciplines themselves rather than their objects. In her reflection, engineering and design are archetypes for a reflection on the conditions not only of creation of new artifacts but of expansion

[23] Chalmers (1999).

of knowledge and new scientific paradigms. For me, it is the starting point for a theory of design as an "in-discipline".

Anne-Françoise Schmid is a philosopher of science and a professor of epistemology who is not only a specialist of the mathematician Poincarré but is also a philosopher, so to speak, "in the field". First, because she has done most of her career at INRA of Lyons, then has participated in multiple research projects in biology, chemistry, and other related disciplines, and has participated in diverse ethical committees to assess scientific practices, for example the evolution of DNA manipulations. Her work is, therefore, centered on radical changes and innovations that she witnesses first hand in the world of sciences.

Anne-Françoise Schmid has developed the concept of "objet intégratif" – that I suggest we translate as "integrative thing" – to reflect primarily on how sciences create something new. She addresses the issue in several texts[24,25] particularly to distinguish the "integrative thing" from complex problems, as defined by Legay,[26] but also to understand contemporary art.[27] As I analyzed one research project after another, the concept of integrative thing seemed to define the research "products" that I observed. These "products", the demonstrators in particular, were strange devices, apart from normal and common usage, and, initially, quite clumsy. We have seen that naming these research products could be a challenge, that narratives came to the rescue and discussing them was also a way to perform them. The concept of "integrative thing" seemed to offer a global way to think about these devices that had properties but (as yet) no identity.[28] I then decided to look at how integrative things originally distribute disciplinary elements in unprecedented ways. The demonstrators in a research project are a case in point. They certainly can be analyzed from their place within the organizations as is the case with "boundary objects",[29] or as "epistemic object",[30] or as the culmination of the technical invention – as described in the "concretization"[31] process. However, I was more interested in them as expansive phenomena that enable inventions "in the sense that neither the history of science, nor the combination of known characteristics, can account for them".[32]

While scientific ideology mostly relies on a discourse of bringing more, or adding more to already known fields, Anne-Françoise Schmid links generativity (defined as the potential to produce something new) to the question of underdetermination, that is subtracting from what is already known. The "integrative

[24] Schmid (2001).

[25] Schmid (2012); Schmid et al. (2011).

[26] Legay and Schmid (2004); Legay (2004).

[27] Mackay (2015).

[28] A very close research is Cetina (1999).

[29] Star (2010).

[30] Ewenstein and Whyte (2009).

[31] Simondon (2001).

[32] Schmid (1998a, b).

thing" is therefore an operation that consists in removing characteristics from an object of knowledge.

> However, this impossible device is in the nature of under-determination. A new device is not a known device + a new characteristic, but an "X" whose characteristics distribute themselves between the disciplines in an unexpected way. To obtain this, one must simply seek what a device is "without" one of its traditional characteristics.[33]

Many contemporary scientific issues are excellent examples of this clash of disciplines. Leo Coutellec, who has worked on Alzheimer "disease",[34] not only shows that no discipline can actually synthesize the whole question but also that all disciplines leave an open unknown. There is a discontinuity that is introduced, that allows us to separate some elements from the disciplinary body of knowledge in order to create a generative space: disciplines are under-determined by each other. Anne-Françoise Schmid also draws attention to the fact that, at the time of the invention, we do not totally know which elements of the disciplines will be displaced or discarded. As I said in the introduction of this chapter, partners in the museum project PLUG had not distributed the tasks nor decided what would change beforehand. More generally, Chap. 5 on design as a composition of tensions gave several examples of conception matrices where designers play with different elements until something appears that makes sense *a posteriori*. The logic of composition is therefore not limited to tangible artifacts, it also applies to disciplines. This is also pointed out by Muriel Mambrini and Armand Hatchuel who both witness scientific evolutions in different fields, the former in genetics and the latter in engineering sciences:

> We will see that the changes of the 'interdisciplinarity logic' are related to the considerable changes within the notion of thing. It is no longer about "shifting" a given thing from one discipline to another and to observe the changes that it undergoes within this journey, but rather to build a thing whose dimensions are themselves disciplinary, that we call 'integrative thing'.

Generative processes are distinct from contiguity or from combinatorial processes:

> Not to view the device as a continuation of a disciplinary reasoning, even compounded, but to connect an X with knowledge islets that are not all prearranged. Thus, the devices are neither obtained by proximity (we are aware of the list of existing devices and propose additional variations), nor by combinatorial algebra (one decouples very different properties to create new devices), but rather by under-determination (what happens to a device if one hypothetically subtracts a property considered as natural?). One is then forced to seek other types of knowledge, skills, to constitute a new thing, an X.[35]

Thus, her proposition works as a methodology to simultaneously observe the creation of new scientific paradigms and their genealogy within and in between the disciplines, and also as a methodology to create new integrative things.

---

[33] Schmid (2012).

[34] Coutellec (2015).

[35] Schmid (2015).

> Another characterization of these devices is that they are not possible realizations, but they first offer an impossible, for which we generate new links between knowledge.[36]

Oxymorons and Porphyry's trees that we studied in Chap. 5 are cases in point. They begin as innovative associations and the interpretative process is so to speak after the fact since it eventually transforms the original concepts to adjust to the new image. The words are deprived of their usual connotations and associations. The meaning-making process indeed consists in finding new and unexpected links between bodies of knowledge that in turn need to evolve.[37]

However, if one agrees on the general process, the integrative thing may take various forms, and its aesthetic appearance is of particular interest to me. In other words, the conceptual space described by Hatchuel et alii, can also be narrative and tangible. For example, at the earlier stages of the research project PLUG, the integrative thing was a premise and a promise: there was a vision with references to fictions. At the end of the project it was a tangible demonstrator performed in the museum. Within this perspective, it became important to see how the disciplines were successful in obtaining an admittedly heterogeneous but consistent thing. The epistemological question had to turn into an aesthetic question.

Engineering sciences and design do not work on what already exists but increase knowledge by increasing the number of artificial devices. "In engineering sciences, we have a mixture between pure laws at a certain scale and the conditions of basic devices that use these laws, which leads to new laws within a new scale".[38] In other words, engineering sciences are not audits of experience or applications of mathematical models.[39] As Anne-Françoise Schmid explains,

> Engineering sciences can no longer meet their interpretation as the application of theories on/in "concrete/actual/practical cases", they are not a particularization and an illustration of the knowledge formed elsewhere but rather the construction of specific order of magnitudes which are necessary both for theories such as data, and knowledge produced for the constitution/formation of the order of magnitude of the foreseen problem, and which, as a result, will be foreseen as partial solutions for comparable orders.[40]

Engineering research or design are simultaneously creating new devices, producing knowledge on these new devices, and reconfiguring existing concepts, methods, and typologies.

From my point of view, the questions are of two orders: about the process of under-determination – which discipline under-determines another – and about the

---

[36] AF Schmid, Proposition d'un cadre théorique pour les objets contemporains in Ibid.

[37] This is consistent with design theory called "CK theory" developed by Hatchuel, Weil and Lemasson. In CK theory, the creation of new expandable concepts is a necessary step to invent something new but it necessarily triggers a reorganization of knowledge or the creation of new knowledge. They call this process "K reordering". Hatchuel, A., Weil, B., et alii, (2003) A new approach of innovative Design: an introduction to CK theory. », in *Proceedings of ICED 03, the 14th International Conference on Engineering Design, Stockholm.*

[38] Guy (2015).

[39] Schmid, *L'âge de l'épistémologie.*

[40] Schmid (1998a, b).

results – what are the research products and how to describe the "demonstrator", the heterogeneous artifact that implements the different scientific constructions. It seemed to me that the demonstrator is more than just a technical end. It projects future effects and uses, proposes narrative instances based on a "what if" plot, that could lead to new knowledge and new devices.[41] In the following section, I want to go back and re-analyze some projects to see how this research and aesthetic experience fits the generative epistemology elaborated by Anne-Françoise Schmid.

7.3.2 *Pluri-Disciplinarity as the Foundation: Challenging the Subordination of Disciplines*

Initially, an engineering research project organizes the subordination of "social" disciplines. At the beginning of the technical developments, social scientists must review social practices that would be affected by the technical invention, and at the end, they organize user tests to ensure the functions and reproducibility of this invention. Within this context, social sciences seem to contribute essentially to the legitimization process. The budgets unevenly allocated to the different disciplines obviously corroborate this understanding. It is within this context *a priori* of human and social sciences' subordination to engineering sciences, that my experience took place. Nonetheless, in letting go of certain disciplinary positions, my experience also reflected another reality: a lack of hierarchy of the disciplines to the extent that they accepted the entire research of a non-defined X device. At first, it should be recognized that an ANR project (a publicly funded French project) can help to build an "inter-disciplinary place" as Muriel Mambrini and Françoise Schmid describe it, that is to say, a place where the "multi-disciplinary" aspects become characteristics of new scientific goals. In the context of engineering research, a form of vertical integration (based on a general model of stratification of systems that I describe earlier in this chapter) gives a shape to this space and opens the possibility of combining multiple disciplines. Giving this possibility does not mean that a flow between disciplines and an expansion of areas of knowledge have taken place. Sometimes, the structure leads to a mere juxtaposition – at best, courteous – of the disciplines. The real challenge for a multidisciplinary team is to go beyond this stratification and juxtaposition and let a space open for new knowledge by under-determining some of their assumptions or concepts, or structuration of facts. It is from this angle that I wanted to analyze a series of narratives and/or tangible "devices", which rely on a breakdown of theories and facts, and borrow from several disciplines through story or image.[42] Among them, the "demonstrator" – as a projective aesthetic device borrowing from several disciplines without achieving a

[41] Gentes (2015a).

[42] Gentes and Selker (2013).

synthesis and articulating fiction and experience – struck me as an unexpected crystallization of elements not limited to a single disciplinary space.

7.3.3 The Aesthetics of the Demonstrator

As seen in Chap. 6, initially, the "demonstrator" pursues a managerial and communicational goal. First, it aggregates different elements and reviews of the research. It makes it possible to test the operating assumptions of the technical invention. Second, its global appearance has to represent results. The demonstrator is part of the project demonstration and communication. However, if we look at it from an aesthetic point of view, it is obvious that the demonstrator is more akin to the Platypus than to the eagle hovering over the peaks. It is made out of heterogeneous elements that are not as yet leveled out or even hidden in a black box that would give the sense of a technical and aesthetic harmony. As we have seen in Chap. 6, it semiotically foregrounds its unfinished and open status. I want to look more closely at this heterogeneity for a moment.

Engineering research organizes, through various writings – scenarios of uses but also documents explaining the technical challenges or "technological nodes" – the mythical story of "problems" to solve.[43] I speak of "mythical stories" to show the purely fictional (but not fictitious) dimension of these problems. For instance, the deployment of a distributed wireless network (between the participants) may use, as a narrative pretext, a place without electricity and without infrastructure that requires people to connect and form spontaneous networks only with the help of their mobile device. This mythical story accompanies the challenge researchers face who make original assumptions: " **if** one took away the stability of a fixed-network infrastructure hardware, would one still have a mobile network?".[44] In all the mobile distributed network projects, the assumption is that a network can be built **without** a central server. On this basis, the research team redefines the overall challenges and creates a demonstrator that will be tested. The demonstrator is fundamentally heterogeneous because it is simultaneously made out of fictions and sciences, of proven technical elements (sometimes even outdated as we saw in Chap. 6) and of hypothetical models, and that it necessarily borrows facts and theories from very different places of knowledge.

It is this heterogeneous device which interests me from an epistemological and generative point of view, because it seems to me that it could be analyzed as an "integrative thing" which enables the invention. To better identify the properties of this integrative thing, I wish now to replay the whole "PLUG" experience cited in the introduction and that has allowed the partners (Institut Telecom, labo CEDRIC du CNAM, Musées des Arts et métiers, game design company: Tetraedge Games)

---

[43] Gentes (2015a).

[44] Méadel et al. (2015).

to invent new forms of interaction and circulations with the data, the devices, and the Museum.[45]

7.4 Under-Determination in Action: Discovering the Generative Properties of the Integrative Thing

I have briefly described the shifts that take place within each of the disciplines involved in an engineering research project to accommodate and carry a bizarre device. No discipline alone can lay claim to its synthesis. I would like to come back to this experience to illustrate my point.

As said previously, the ANR-RIAM PLUG project involved several disciplines and fields of expertise: computer sciences, information and communication sciences, museology and research in game design. They were each involved in establishing a new relationship with the Museum of Arts and Crafts. The project was a place of interdisciplinarity and the theater of several challenges and concerns: a technical challenge (mobile distributed network); a Human-Machine Interaction (HMI) challenge (including a research for new gestures, new menus, learning and attention theories, and more generally of play) and a cultural production challenge (narrative approaches, game play, etc.). The exemplarity of PLUG is that none of the disciplines involved was left untouched. It is now my understanding that such a research relied on a series of under-determinations between the disciplines that I explain in the next sections.[46]

7.4.1 Computer Sciences' Under-Determination of Game Design Theories

The project challenged one of the foundations of game design, specifically in the way games create a "pure space ",[47] "a Magic Circle", in other words a closed session that is simultaneously physical, symbolic and imaginary, within which the player is fully involved. The effectiveness of game design relies on its flow[48] theory and on its screen interaction models. Contrariwise, our pervasive application scattered the attention between different "media" and channels. In effect, tangible artifacts were integrated within the virtual game through sensors installed on both the mobile device and the artifacts. The latter would appear on the interface of the game's mobile device when the player moved closer. The very principle of pervasive

[45] 2007–2009: Projet RIAM, « PLUG, Play Ubiquitous Games and Play more » (Jeux pervasifs).

[46] For other related work on that topic, see Gentes (2015b).

[47] Caillois (1992).

[48] Csikszentmihalyi (1991).

computing, which is to be included in all kinds of objects and locations and to distribute computing beyond conventional terminals (computers, phones, tablets), challenged the "flow" of the player because the aim was to switch the player's focus from the "terminal/device" to the museum and vice versa. In hindsight, the research could be summed up by a question: how to forego this "Magic Circle", and nevertheless provide for a game?[49]

To explore forms of interactive visits between the virtual and real and physical worlds, game design research was going to have to borrow the notions of context and situated action from the social and computer interaction theories of "pervasive computing".[50] Game design researchers had to integrate the way in which people interact with their environment in both direct and mediated ways.[51]

This under-determination of game design theory by pervasive computer sciences first seemed to reduce the effectiveness of the game design discipline. But at the same time, the demonstrator created new forms of interaction between users, the world, and the machine. In particular, it led game design theorists onto concepts developed in literature such as "the uncanny". The latter concept, akin to the feeling of "déjà vu", was then elaborated to explain how it was a specific form that captured the players' attention.[52] The uncanny and its application to game theory made it possible to challenge the logic of magic circle and to define a form of scattered attention. In other words, pervasive computing under-determined the discipline of game design and defined original concepts that changed both disciplines while introducing humanities through a literary format: fantastic literature and its use of the uncanny.

7.4.2 *Museology and Human Sciences' Under-Determination of Computer Sciences*

Computer and network sciences were destabilized because the Museum refused to use GPS systems or Wi-Fi to navigate during the visit. Researchers in museology pointed out the ethos of the relation with visitors and were skeptical about using yet another sophisticated expensive artifact to "augment the visit" by tracing what visitors did. Eventually, the stable mode of operation, that is linked to a centralized architecture on which the overall reliability and safety of the exchanges of information relies on, was put into question. For computer scientists, the issue was how to organize communication between users even though the network would no longer organize the conservation and the data distribution from a central server. In other

[49] Simatic, Astic, Aunis, Gentes, et alii (2009).

[50] Nieuwdorp (2007).

[51] Gaste and Gentes (2013).

[52] Gentes and Mollon (2015).

words, two fundamental resources: the memory and the dissemination/diffusion that are supported by centralized computer systems were compromised.

This challenge was partially settled by considering temporary modes of data conservation within RFID tags that were placed next to the devices and within the NFC phones ("near field communication" that can read the RFID tags) and exploit/use the "reading" and "writing" potential of these tags. But it was also settled by redefining what a visitor is and what relation she entertains with the museum.

In fact, "visitors" were the solution to the technical problem by physically moving to deliver the information from one point to another within the building of the museum: like bees they would pollinate the artifacts. The participants had to find the right information to bring it to the right museum artifacts. The testers understood very adequately that they themselves constituted the network architecture through their whereabouts. Both the notions of architecture and the notion of visitors were redefined by the under-determination of computer science by the museology and vice-versa. The role of the visitor, his/her place within the space, his/her ability to establish a site visit, undermined a vision of computing that had to manage everything: time, memory, and mobility. An architecture was no longer a fixed organization relying on non-living actants but a living mechanism based on people. However, the visitor was not only a person but a physical component of a computer system architecture.

7.4.3 Human-Machine Interfaces' Under-Determination of Information and Communication Sciences

Finally, information and communication sciences tenets were jeopardized by human-computer interaction (HCI). Information and communication sciences do not consider communication as a mechanical movement or storage of information bits but rather as a social and semiotic gesture. This standpoint opens up, among other things, the possibility to define certain interactions between mankind and machines as a reading practice. Within the PLUG project, the very fact of "catching" a few text snippets to "drop them off" near another device did not resemble what is traditionally defined as reading and writing. However, the discipline had to consider the actions of the equipped visitors and finally redefine the articulation between the gestures and technologies not as interaction but as a form of "proto reading" and "proto writing."

In effect, the practices supported by the technical and semiotic apparatus showed that the system endorsed the user as a reader inasmuch as she constituted a documentary space, connected different texts, judged the relevance of the representations. This new reading and writing model that incorporated a new gesture and a form of minimum reading and writing, was triggered by the under-determination of information and communication sciences by computer science. It also enabled us to redefine the player's task/role and the functional modalities of the RFID tags.

Describing the user activity on the device/terminal in terms of proto-writing and proto-reading transformed the vision of a user-activator of technology accomplishing a task to a co-developer of meaning in a situation. In addition, it made it possible to take a closer look at the tenuous forms of reading and writing often neglected within information and communication sciences.

7.4.4 The Demonstrator as an Integrative Thing Between the Disciplines

At the end of the research, the integrative thing is not only a premise and a promise. It is embodied in a new tangible device, that no discipline can synthesize: the demonstrator. Within the PLUG framework, the design was innovative in the sense that a whole set of concepts was re-qualified by a device that was, as a result, irreducible to any classical categorization. The PLUG artifact was thus very different from the product of an alternative concept-design – which I would describe as non-innovative: multimedia audio guides. In the latter case, one adds a modality (the image) to a media (the audio guide) without questioning the forms of visit, the role of the museum within the editorial enunciation, and of course without bringing new knowledge in mobile distributed networks nor challenging video games theories. In addition, multimedia audio guides stick to predictable activities while in PLUG the conventional figures of the user (as a partner of the technical system) /visitor (as a partner of the museum institution)/ player (as a driving force of fiction) were redefined.

The process of multiple under-determination that led to this integrative thing was not planned. The participants did not know in advance what to give or what to take. However, the whole process showed several characteristics. First, the project was defined as a research question not as the resolution of a problem. We had to produce an X. The way the question was asked without prior determination, led the disciplines to look for new articulations between their areas of competence. Second, there was no hierarchy of disciplines but a levelling of their contributions. Each discipline made a kind of inventory of what it knew or at least laid out what the question suggested, to be redistributed to the others. Finally, the composition of different elements of disciplinary knowledge meant that each discipline considered what would be its object (conceptual or tangible) if some of its characteristics were removed from it. To come back to Anne-Françoise Schmid's definition, the integrative thing is first a "device" or discipline from which some properties are taken away.

The fabricated device was therefore not an augmented device (even if we speak of augmented reality) but a strange device generated by the under-determination of each discipline. I think that this under-determination is also represented in some fictions that start the projects for example when the scenario of use describes an entire telecommunication infrastructure going down and thus justifies the research for an alternative. I find interesting that some of the scenarios have within their script a representation of the epistemological operation that consists in removing characteristics from devices in order to find new ones.

7.5 The Forgotten Discipline: The Humanities

The epistemology of invention is therefore dependent on a definition of science as expansive and produced by under-determination. We can look at it from a philosophical, logical, and even formal perspective. But we can also look at it, as I have done in this book, from an aesthetic perspective. What I mean is that the concepts developed by the humanities like fiction, aesthetic experience, representation, and media, are powerful instruments to explain the expansive process. The next two sections are dedicated to a last analysis of PLUG and the autonomous aesthetic plane that it created. I also want to give a summary of the different chapters so as to emphasize "the humanities in action". For me, the humanities not only bring methods of analysis of design practices but belong to the episteme of design in its expansive capacity.

7.5.1 Science as a Generative Space of Fiction and Experience

PLUG planned a new technical and cultural experience in a museum. However, the project's description was not shaped by a technical research question nor by a specific social goal such as learning or collaborating. It nevertheless identified an unknown zone qualified by a certain number of properties and visions that I want to analyze here. In other words, despite its "undecidability", the proposal was aesthetically structured.

First, the scientific project included fiction. The project's vision was supported by a reference to a science fiction book and was mentioned in the response to the bid:

> Today the device/terminal available to the player for a ubiquitous game does not come straight off-the-shelf, it would be tantamount to "pussyfooting" (The Age of Pussyfoot (1969).[53]

Eric Gressier-Soudan often quoted this particular reference that he shared with the team. The book is famous within the computer science community (an article by Luca Cardelli from Microsoft Research demonstrates the impact of the book on researchers' culture[54]) because the author is one of the rare who anticipated the information age particularly through a "joymaker":

---

[53] RIAM PLUG – document for the call.

[54] http://lucacardelli.name/indexExtra.html

> *The remote-access computer transponder called the "joymaker" is your most valuable single possession in your new life. If you can imagine a combination of telephone, credit card, alarm clock, pocket bar, reference library, and full-time secretary, you will have sketched some of the functions provided by your joymaker.*[55]

This vision was complemented by other references particularly from the J.K. Rowling's series: Harry Potter.[56] The game designer referred to the "living" portraits who talk and move when one of the characters enters a room (they are the keepers of the "houses") and more generally when there is some agitation in a place. This feature of the novels fueled the notion that artifacts ought to come alive and deliver messages when a visitor passes by. Inspired by the living paintings, the team switched to "living artifacts". These references not only triggered the memory of a specific item (or prop in the works of fiction) they offered (as we have seen in more detail in Chap. 4) a whole world where the technical invention fits with activities, atmosphere, characters' personality, etc. Fiction here plays a role of anticipation for the research goals. It defines indirectly what is the expected experience.

I want to emphasize the word "indirectly". Of course, fiction is a plane of its own that does not represent "reality". But readers still have to see the articulations between actants, actions, worlds, and consider them as meaningful within the story, as well as meaningful in relation to their own experience. The fictitious world has both an internal coherence and a pragmatic relevance to the reader.[57] In addition, fiction also brings its own aesthetics: its own language, images, descriptions. The research project is therefore nourished by the potential of the fiction based on the story and actions and supported by its aesthetic cues. Fiction is meaningful in its own way but it also opens up a horizon of possible things to live and experience.

I also want to insist on the word "experience". For me the word refers to two distinct things. First, the word is generally understood as a set of cognitive and sensitive properties related to a user and a situation. PLUG's project manager defined the technical artifact indirectly by describing some properties related to its potential users and their environment. He explained that we would create a portable device that would interact with other artifacts and other users:

> Two features are essential to this apparatus: the terminal is a personal object like a mobile phone; it accompanies the player in all places at all times. Furthermore, it is in constant communication with its surrounding environment (buildings, urban props, consumer goods, works of art or exhibited objects, services, other players…)[58]

In the texts laying out the project foundations, what was described was the overall experience of the user: her relation to the device is "personal"; the device is like a life assistant as it is continuously by her side; the situation is made of a pervasive environment that immerses each user in a reactive space. The vision was particularly original because the user could play with or could be played by the system:

[55] Frederik Pohl, *The Age of the Pussyfoot* (Ballantine, 1969).

[56] Harfield (2008).

[57] Harris (2000).

[58] RIAM PLUG Project document for the call.

> the player can be in the center of a technological device that immerses him- or herself in the game and its atmosphere, but also the game-design can trick him/her.[59]

The project manager therefore also introduced an uncertainty for the user who could be tricked, or surprised by the device. The usual engineering scenario where the user is the master of the system and controls her equipment gave way to a more interesting figure of the user/used and a whole array of unexpected interactions. I want to emphasize again that the texts did not describe the object proper but focused on the general experience that the authors wanted to emulate. So to speak, the experience was delineated from the outside, as the negative in a photography.

That is why I suggest that the research project acts as a translation of the fiction. Here, elaborating on translation studies, I use the word experience with a slightly different meaning because I focus on the operation of transformation and on the situation of utterance. It is a well-known goal of translation to provide readers of the translation with the same experience as in the original text. The translator's main responsibility is to make a text readable to an audience with a different language. Her work consists in choosing the best possible equivalences of meaning. But in translation theories, there are at least two focuses. The first one emphasizes the fidelity to the first text and the best equivalence means that the translator has to recreate the cultural background of the first text and has to stick to the writer's horizon of writing. The second focus is on the situation of reception. The translation is therefore part of a strategy of communication that considers the context of utterance, to build on the vocabulary and more generally the language skills of what "skopos theory" calls the "target readers".[60] In this functional theory, each work is reinterpreted and reassessed from the standpoint of different horizons of expectations of readers – interpreters. "Now, instead of equivalence of meaning, many authors say "functional equivalence or skopos theory: "a translation (especially in the case of texts with an aesthetic purpose) must produce the same effect as that of the original. This is called equal exchange value, which becomes a negotiable entity."[61] Within this paradigm, to translate is not to use a dictionary of equivalence, but each time to elucidate the contexts of utterance. Translation is not about finding the one and only answer to a problem of cultural transposition but rather to adapt the response to the concrete situation of communication. It is thus understood that the space of translation is a flexible space, negotiated between interlocutors. This is what Andrew Chesterman[62] after Gideon Toury,[63] called the "relation norm" that implies that the translator finds the best option for the sake of similarity but without predefined equivalence. By abandoning the idea of a strong equivalence in favor of a negotiated production between the original text and the actual situation of its release, it also gives translators a real role of mediators, who not only support the

[59] RIAM PLUG Project document for the call.

[60] Reiss and Vermeer (2013).

[61] Eco (2004).

[62] Olohan (2000).

[63] Toury (1995).

text source but care about its re-appropriation. Translation is in effect rhetorical and not only linguistic. The emphasis is on the actual present situation to which the translator brings elements from the first "offer of information".[64] I suggest that the researchers act as translators when they strive to give the same impressions to the "target readers" as in the first texts (the fictions). The target audience (in our case the potential users of PLUG) should experience something that makes sense within their own world, culture, perspective. This is a different emphasis on experience. Here it is the performative virtues of a new text/artifact that can always refer to a first original text. The latter can be interpreted again with different artifacts, words, activities. The PLUG team, therefore, transformed the first offer of information into a second offer of information that kept some of the properties of the first text. This kind of framing that I like to call "fiction led research" can be particularly powerful as it focuses on the richness of experiences in situation rather than only a technical achievement.

As a matter of fact, the properties that were stated in the research prefiguration of PLUG, not only led to a game but also to an interesting technical innovation thanks to Michel Simatic who developed the concept and technology of "vector clock".[65] This technical invention which contributed to the content synchronization of the distributed mobile system, finally was at the technical basis of our project. However, this invention was not pre-scripted at the beginning of the project, but it was foretold by the narratives and the narratives were translated into an experience that deliberately emulated the experience born from the narratives.

7.5.2 Lessons Learned Along the Way

Social sciences and human sciences are intertwined in terms of their objectives and methods. However, I have maintained the distinction between the two different approaches for the following reasons. On the one hand, I think that the usefulness of social sciences in design no longer needs to be proven. The literature on the use of ethnography, sociology, and such disciplines in the design process is huge. On the other hand, I think that the humanities have not received as much scientific recognition in general as in their role in the invention. I think that we need to consider the contribution of the humanities to design not only because they help us understand design activities but also because they are part of the design episteme.

The different chapters of this book have tried to reassert the role of the humanities in research and design practice and to elaborate a theoretical framework to do so. The goal is to look at the autonomous plane that invention needs – a plane made of names, fictions, figures of speech, narratives, debates – before it is adjusted to the existing practices. In the field where I come from – information and communication

[64] Reiss et Vermeer, *Towards a General Theory of Translational Action*.

[65] Simatic and Gentes (2009).

sciences – we try to link these two “matrices”.[66] We consider that it is particularly relevant to consider media and new media from both social sciences and the humanities. However, even within a field that was, *a priori*, founded on the combination of these two traditions, this association is not so straightforward. The aim here was not to revisit the difficulties of the disciplines’ legitimacy but to emphasize that we are dealing with issues of granularity and the linkage of complex levels of analyses between semiotics and social organizations.

But first, we need to go back and question Foucault’s definition. We saw in the Chapter of introduction that, in *The Order of Things,* Foucault proposes a definition of the “humanities”:

> that region where the laws and forms of a language hold sway, but where, nevertheless, they remain on the edge of themselves, enabling man to introduce into them the play of his representations, in that region arise the study of literature and myths, the analysis of all oral expressions and written documents, in short, the analysis of the verbal traces that a culture or an individual may leave behind them.[67]

I want to elaborate on Foucault’s definition by pointing out that it is not only a question of linguistics and representation but also a question of communication and media. The notion of representation tends to limit the debate on mimesis. Representation means that the sign has a semiotically mimetic relation to what it describes. First, we need to understand that the poetic dimension of languages (either linguistic or visual) obeys other rules than strictly representative ones. This autonomous plane of language is what allows the “play” within languages, in other words their creativity. Second, there is no disembodied language. We are not pure thoughts. The concepts of communication and media open a more autonomous place and look at how humans play with materials, time and space that detach them from the here and now of action: “Distantiation creates the possibility of media, which become both means and ends in themselves”.[68] Mediation to power (through money for instance), mediation to death (through religious art for instance) are not only ways to “represent” actions or worldviews. They have their own aesthetics, their own history, their own relations of intertextuality and intermediality. Mediation also addresses the more general use of a media as a plane of expression independent of any positivist ideology of representation of the “real world”. Foucault’s definition of philology seems to me to concentrate on the “message” leaving aside the fact that the “medium is the message” as stated by Marshal McLuhan.[69] Media are not summed up by the issue of “representation”. They are part of a tangible expression as well as part of a process of communication. As stated by Guillory:

> Grasping the nature of mediation depends in my view rather on affirming the communicative function in social relations, that is, **the possibility of communication**.[70]

---

[66] Souchier et al. (2003).

[67] Foucault (1966).

[68] Guillory (2010).

[69] McLuhan (1965).

[70] Guillory, « Genesis of the Media Concept ».

If we want to study design/conception, we need to understand how it explores all the poetic strategies of virtual distantiation within a perspective of communication. All artificial objects are a way to distance ourselves from the world whether or not the final products are information technologies. In other words, looking at design from the humanities perspective means that we consider this process of distantiation that breaks free from social or technical determinisms not only through personal expression but within a social horizon of communication, hence the definition of things as fundamentally debatable.

I started this book with an analysis of information and communication technologies for a good reason. I needed to understand how they are used in our contemporary culture, but more importantly, from a design perspective, I analyzed them as objects that create a meaningful generative space both autonomous and dependent on human activities. Indeed, they are not only a specific type of artifacts and part of the ecology of our contemporary system (the society of information and communication) they also provide an aesthetic plan of exploration and realization. The second chapter therefore considered how these technologies are emblematic of what I suggest to call the reflexivity of design. Design/conception uses media to step back and test diverse possibilities. Medium and media are here considered from the point of view of the material agency of these tangible artifacts but also from the point of view of their expressive and communicative properties. Form does not "follow" function but "explores" functions. In this respect, the humanities give us a special insight on what characterizes media, in terms of their formal, cultural and social properties. All issues of knowledge formatting and representation, introduction and legitimacy of actors, means of communication, archiving and the diversity of memory formations, the power of data are all issues connecting media and design.

In Chap. 3, I showed how theories of invention rely on a model of action whereby effectiveness, efficiency, speed, learnability, etc. are the focus of design. Social sciences have developed this model and helped designers take into consideration the physical as well as the cultural and psychological points of view of the users. It means that designing always embed a physical and psychological model of human uses. But we need to address a third dimension of humans' activity which is their intellectual activity, the way that humans represent and share their worldviews. The user is therefore not only a multitasker and an aesthete, she is also somebody who works on symbols. I think that it is fundamental to take this latter figure seriously as it has been mostly disregarded as "unreliable" in terms of engineering.[71] Even more importantly in this demonstration, I have tried to show that we can multiply figures of the users. In other words, the "users" are useful not only to ground the invention in a predictable social activity, but also because they are not "real" users but figures that move in an imaginary space that help us to focus on different properties of a foreseeable system. They belong to a plane of representation and mediation that is at the same time connected to and independent from activities. Firstly, it is connected in the sense that human experience shapes it but also because it shapes the way humans think and communicate about their experience. Secondly, it is

[71] Except in all the creative industries.

independent in the sense that there is no "cause to consequence" modality, but a freedom to explore the different aspects of users/characters and how they appear on and are shaped by different media.

Chapter 4 dealt with the plane of language as a poetical space that not only tries to represent projects and inventions but also that values the connotations of words and images for themselves. The creators of words enter a paradigm where they open the meaning of their artifact by associating it with linguistic references, images, and sounds. This poetic space has a certain meaning today that does not foreclose potential future interpretations. It is therefore generic and expansive for several reasons. First, because these poetic creations detach the interpretation from a "pure" activity. The effort to name is part of the effort to go beyond the sheer imposition of forces and to enter the world of culture. Second, because they play between different media: linguistic, images, or tangible elements. Every plane is in a position of possible under-determination of the other so that the meaning of one semiotic composition is always somehow challenged by other semiotic configurations. Finally, their interpretation changes over time. Not only are we getting used to these technologies through a slow process of adoption but we can change the meaning of the words meant to represent them and consequently change the way we think about them and use them. The history and destiny of the machines are related but not totally dependent on the history of words and connotations that represent them. Whether they are names, logos, or narratives, their function is to foretell the future in what we can call an "expansive literature".

It was then necessary to consider how designers, engineers, and artists bring together these different elements in a matrix where unforeseen, unplanned connections take place. There are two levels to this discussion. First, professional designers cannot leave these unforeseen encounters to chance. They have to organize a field of tensions between different meaning making systems for instance through comparing corpuses or through using different "architexts". Somehow, the goal is not to control tools to accomplish a certain task defined by a problem but to play with instruments that give different performances, different versions, preventing one from fixing a meaning over another. Chapter 5 therefore focused on the spatial plane of composition as a complementary vision of the design project. What matters in a composition is how a field of tensions is created and elements manipulated. Composition here is defined as a matrix that brings together heterogeneous elements. The second part of the discussion focused more on the details of these encounters in particular how they are based on what CK theory qualifies as a "crazy concept" and that I analyzed as an oxymoron. The oxymoron helps us understand how, first, the association is mind boggling but forces a process of interpretation that not only considers the seemingly impossible conjunction and brings other associations, but redefines the initial terms too. This metaphoric process reconfigures the primary knowledge bases. The whole elaboration is therefore about discordances and harmonies by re-adjustment of the whole tree of significations. I found the porphyrian tree as it is discussed by Eco interesting not because it relies on a static vision of knowledge (which it did in the Aristotelian tradition) but because it shows how the oxymora reconfigure the elements of language or image in a new composition that finally makes (new) sense. In other words, this chapter tried to show how two seemingly divergent creative activities, one of deconstruction, the other of composition, can be regarded as a poetic effort to create new forms of coherence.

This field of tensions is supported as well as it triggers another important element of the humanities of design. The "unknown", whether it is an unusual turn of phrase or a new artifact, cannot be named. Chapter 6 echoes the demonstration of Chap. 3 in that it analyzes how the invention struggles to find founding concepts. From the point of view of design, it means that creators focus on "things" that are debated rather than "objects" that are used. I want to stress two aspects. First, considering "things" means that we refuse two epistemological standpoints. We refuse the ontology of artifacts that would already be given and that the analysis would discover. We also refuse the pure subjective point of view whereby everybody has a personal sensitive experience that cannot be shared. The standpoint on "things" is communicative in the sense of American pragmatism (Dewey) and the Frankfurt school (Habermas). In that perspective, a thing can be debated with rational arguments that question the reasons, and the individual is defined by her capacity of communicative action. Second, debates are a generative space and not only a space of controversy. They are not only a space of confrontation of pros and cons. Debates discuss definitions and come up with new concepts because they have to work with non-knowledge. This point is fundamental as it leads to our epistemological reflection. The questioning and expansive nature of things is both related to their conceptualization but also to the fact that they are remediated in multiple situations and audiences. Their inconclusiveness is not just linguistic (in other word it is not a question of solving a stated problem) it is media related. The plan of "things" is a plan of multiple media: objects, discourses, images, activities. The generation of new interpretations is directly related to the clash between the different media, and the non-decidability of the trans media experience depending on different semiotic configurations. However, some things can be perceived as more opened than others. In fact, certain objects, discourses, and mediation keep more un-decidability than others in relation to specific situations and audiences. Their "strangeness" can of course be fortuitous, but it can also be carefully maintained within an explicit design perspective.

In these chapters dealing with how design is about media, figures, naming, composition, and debate, the pluri-disciplinarity is not about adding disciplines to better deal with the final object and her user, but about the process of conception itself. Seen through the lens of the humanities, conception is a plane that is autonomous while connected in certain ways with social experience. This plane is generic because not only the different elements of the design situation but also the different disciplines are freed from their epistemological determinism and can therefore under-determine each other.

7.6 Conclusion: Contriving Observation and Analysis of the In-Discipline of Design

Now I want to confess that it took me a while to be able to understand research projects from this standpoint. I needed to let myself be destabilized in order to produce new objects of research and what I hope are original perspectives on design. Far from resting on a stable body of knowledge, I needed to agree to be challenged by other disciplines. In the process, my discipline, information and communication sciences, was under-determined by computer science, game design, museology, literature, art, etc. To be part of the in-discipline of design meant changing my usual scientific posture, methodologies, and predetermined disciplinary concepts, and in particular to rely on multiple disciplines so as to deconstruct my own and expand it. A critical stance is a good departure point to question one's ideology but it is not enough to think of and produce alternative metaphors, things, services, artifacts. In any event, I found it extremely difficult to extricate myself from multiple scientific legitimate discourses that came either from engineering sciences, or social sciences. It was hard to put humanities into practice in an engineering research institution. It was difficult to have fellow social scientists admit that I could actually "make things" within a social science department. And people from the humanities did not see how they could learn from engineering practices. However, I am not alone in this endeavor as the references in this book show. And as the saying goes, "the proof of the pudding is in the eating": colleagues and fellow scientists around me were generous enough to give me the freedom to exercise my indiscipline. Now I want to show the overall benefits of choosing this path.

7.6.1 The Methodological Benefit of the Integrative Thing

The concept of "integrative thing" allowed me to get rid of the dominant discourse of engineering sciences, as well as to put forward the question of heterogeneous systems that nonetheless achieve a form of coherence.

First, it contests the dominant ideology of engineering research as a research for "more": "more" bandwidth, "more" pixels, "more" speed, "more" mobility, "more" power, "more" miniaturization (Moore's law is repeatedly cited within the ICT field). The discourse that accompanies engineering invention is a speech of power and gain. It is a scientific ideology that presents the invention as an addition and not as a difference. The discourse on the "plus" creates expectations and it traps researchers trying to qualify (and quantify) contemporary engineering. I find it is extremely difficult to think outside of this prism because it introduces a bias that prevents the deployment of the right methods of observation and interpretation. Contrariwise, using the concept of an integrative thing allows us to examine what the disciplines abandon in order to be generative, or how they make room for a new device. The concept of an integrative thing has thus a methodological scope because it organizes the observation in terms of what was removed within the disciplines.

Secondly, the integrative thing – like a demonstrator – provides a form of coherence. A demonstrator in a research project is made of heterogeneous elements that are not synthesizable by a single discipline. It is this X that the research builds and that is the result of an under-determination of disciplines. Yet, it also composes these heterogeneous elements. In the different experiments, consistency is linked to the redefinition of concepts such as for example "game", "writing" and "reading", and "pervasive computing" by each discipline. The integrative thing thus re-distributes characteristics of the situation, user, service, and technical infrastructure in order to create a coherent system. Undoubtedly, the forms and operations of coherence vary from one research project to another but they are present when the project technically works and is "performed" by users.

The challenge is to go from a teleological (or functional) and critical (interpretative) analysis of devices in research; to a conceptive analysis: we need to understand how the processes and productions are truly generative through a certain number of operations: like removing one property of an "object" to invent a new one, but also like composition, narration, naming, debating that were presented along the book.

7.6.2 The In-Discipline of Design

To come back to the claim of design practitioners, we have seen that not only do they use multiple disciplines to create an X, they actually organize a deconstruction of disciplines. Designing is fundamentally multidisciplinary because it creates new things in and between disciplines that transform their original concepts and methods. It is this deconstruction and composition of disciplines, this generative plane and dynamics in between disciplines, that I finally chose to call the "in-discipline of design". Before choosing this expression, a few others were suggested that I eventually rejected but that I would like to mention since they brought interesting insights in the analysis of design.

George Amar, former head of the direction of prospective research at RATP, influenced by Duchamp who coined the expression "infra-mince" suggested to use: "infra discipline". I found it very compelling for several reasons. Amongst the small pieces of papers that the avant-garde artist wrote, there is one that is called "infra-mince: infra thin[72]" (Fig. 7.2).

His entire career, Duchamp tried to capture this infra-thin event and space that goes from potential to actual. He was not so much interested in the final results as in the process holding the promise of something else. For instance, "3 stoppages étalon" ("3 Standard Stoppages" 1913–1914/1964) is a work of art that takes a rope and measures the mètre étalon (the "Standard Meter Bar") then throws the rope on the ground then keeps three iterations of the results. As George Amar pointed out,

[72] Dation, 1997, Numéro d'inventaire: AM 1997–98 (1), translation is mine.

Le possible est
un inframince
La possibilité de plusieurs

tubes de couleur
de devenir un Seurat est
« l'explication » concrète
du possible comme infra
mince.
Le possible impliquant
le devenir – le passage de
l'un à l'autre a lieu
dans l'infra mince.
allégorie sur l'« oubli »

The possible is
an infra-thin
the possibility of several

paint tubes
to become a Seurat's painting is
the concrete "explanation"
of the possible as infra
thin
The possible involving
The becoming – the passage from
One to another takes place
In the infra-thin.
Allegory for the forgotten

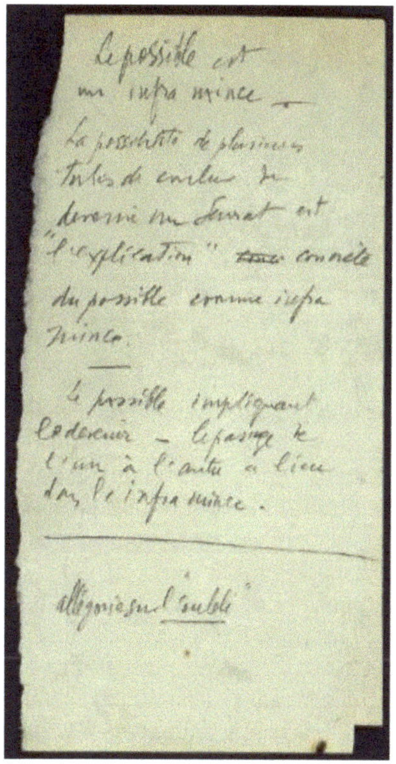

Fig. 7.2 Marcel Duchamp, "inframince" © Jean-Claude Planchet – Centre Pompidou, MNAM-CCI /Dist. RMN-GP © The estate of Marcel Duchamp/Adagp, Paris

more radical artists (like Dada for instance) would probably have broken the "Standard Meter Bar" or a facsimile. But Duchamp kept the tension between the metrics and a new formulation resulting from a series of gestures, certain rules (keeping the Standard Meter as a measurement tool), and chance. The reason George Amar suggested that infra-discipline would suit my purpose was because it described the under-current that changes one thing into another, the dynamic as well as the locus of conception. Similarly, the French Oulipo[73] writer, Georges Perec, spoke of "infra-ordinary"[74]: like the invisible yet sustaining fabric of life against which we pit the extraordinary. Thus, both artists were not trying to define what is ordinary or extra-ordinary, but they trained their gaze to capture the smallest changes. However, these artists did not describe an anthropological method that would pick up every detail, but an aesthetic space of expansion. New interrelations could be drawn. New compositions could be made from these observations.

Another way to analyze an expansive and dynamic plane between the disciplines, is to see that the focus is not on the objects nor "actants" but the relation between actants. As Fluxus artist, Dick Higgins, put forward, art should explore the relation between "painting and shoes".[75] More generally, Fluxus artists wanted to explore the "inter-medium". They no longer cared about the history of art, and the careful distinctions between artistic disciplines (painting, sculpture, or music, for instance). They came up with the concept of inter-mediality, a concept that since then has had a huge success in the humanities, precisely because it depicts the way artists no longer care about disciplines nor are their works built along the lines of a simple medium. It has been useful to understand the clash between disciplines but also the way that an artistic object may be presented in multiple venues, with multiple re-arrangements, hence varying over time and space. The focus therefore is no longer about frontiers between disciplines, but about the changes of an object through an iteration of situations. The notion of inter-discipline might therefore capture not only the gesture of getting rid of traditional disciplines but also of capturing the contextual, situated, changes.

Still, I prefer the expression "in-discipline". First for political reasons. Design comes at a price both personal and institutional. The gathering of disciplines can be conflictual and one needs to carefully study how art and research institutions, and academic organizations, tackle the issue for individuals. I have alluded to the fact that certain projects organize the meeting of disciplines and, while not necessarily guaranteeing expansion, they are still a deliberate and worthy attempt to create a generative space between disciplines.

I also want to keep the concept of in-discipline because a design space questions frontiers and challenges territories. It is a war machine against other war machines. Here, I want to make clear that design/practice, as a body of knowledge, practices, objects, and the way they are taught in schools and exercised by professionals, is of course with the same limits as other disciplines. It builds and defends a territory. However, design as conception (design/conception), that can be observed in different situations that do not always involve designers, is another matter. The under-determination of disciplines does not depend on one discipline only, even though design/practice as a discipline is emblematic of the process of under-determination.

[73] OULIPO Ouvroir de Littérature Potentielle, a literary movement that used strict unusual conceptive rules to invent new ways of writing.

[74] Perec (2008).

[75] Higgins and Higgins (2001).

I would argue that design/conception is an in-discipline and that it manifests itself strongly in design/practice. Hence the possible confusion between the two. A theory of design that takes into consideration the conceptive characteristics of research activities therefore challenges a vision of science that derives "naturally" from its object. Conversely, it emphasizes how disciplines build, create, imagine, their object.

Moreover, the expression was used by the philosopher of art and design Jacques Rancière to define both the evasion from the borders of disciplines and the aesthetics of knowledge. As he points out:

> A discipline, in effect, is not first of all the definition of a set of methods appropriate to a certain domain or a certain type of object. It is first the very constitution of this object as an object of thought, the demonstration of a certain idea of knowledge – in other words, a certain idea of the rapport between knowledge and a distribution of positions.[76]

Bourdieu's theses on art[77] consider that there are only two types of knowledge and non-knowledge: the know-how (savoir faire) and the knowledge of one's position, or the social distribution of roles the "know-who". These two knowledges (savoirs) are related to a logic of ends. There is indeed a way to look at the research, design, and art examples in this book and to emphasize the asymmetry of disciplines and actors. However, I hope to have shown that such a vision would be partial and beside the point. There is more to the research activity than just a plan of optimization of means towards an end, and a social distribution of roles. I hope to have shown that naming, discussing, and composing, are an end to themselves because they are part of the generative process. Elaborating on Kant's Critique of Judgment, Rancière puts forward the fact that the aesthetic experience looks at forms without a "goal":

> The aesthetic gaze which sees the form of the palace is without relation with its functional perfection, and with its inscription in an order of society. It acts as if the gaze could be detached from the double rapport of the palace with the knowledge [savoir] invested in its fabrication, and the knowledge [savoir] of the social order which provides it with its context.[78]

The aesthetic experience therefore disjoints knowledge on how to make and who to be. In the aesthetic experience, the know-how and the know-who can be suspended. Something that is therefore neither determined by both types of knowledge can take place. What happens then is of course dependent on education and general social position as was demonstrated by Bourdieu, but is not limited along those terms. I want to emphasize that this book, rather than allocating power to actors or actants, or tracing how means are aligned and strategically used, tried to explore the aesthetics of research and invention.

If we go back to the "figures of the user" that I describe in Chap. 3, we can see how this is at play. Design/practice certainly needs a vision of the user as an acting

---

[76] Rancière (2006).

[77] Bourdieu (1986).

[78] Rancière, « Thinking between disciplines ».

figure who acts strategically. The use of social science in design therefore helps in assessing this particular figure of the user and more generally to envision the continuum between the technical core of the machine and social activities. But design/practice also needs a perception of the aesthetics of the machine that does not depend on a strict alignment of means to goals. The continuum between the technical object and social activities is seen through the lens of a diversity of aesthetic experiences that vary in time and space, and more generally depend on situations. These situations are contexts that, together with the presence of the different actors, build the meaning of the new thing. The use of humanities in design therefore helps in composing the different attributes of a meaningful context. The figure here is that of an aesthete and meaning making interpreter of things. As Rancière says:

> There are spectacles which disassociate the gaze from the hand and transform the worker into an aesthete.[79]

To enter the in-discipline of design is therefore to look for new ways to relate things and disciplines together.

This perspective of the in-discipline of design/conception is close to critical engineering discussed in Chap. 6, and, for different reasons, "design theory" that we have regularly quoted in this book.[80] When the latter defines a design question as different from the resolution of a problem, and how to come up with an innovative concept, an X that is neither true nor false, it introduces a plane that severs the usual links of causality, as well as the know-how and the know-who. While "design theory" does not describe this place as the in-discipline of design (probably to avoid some backlash from the discipline of engineering design), it still challenges causal relationships and demonstrates the inevitable creation of a "crazy" conceptual space.

To sum up, there are many ways to look at how the humanities help us think about design/conception as an in-discipline:

- They contribute to a definition of the designed objects: the humanities are necessary because they look at the designed object made of semiotic, narrative, and inter-medial meaning making properties.
- They support methodological considerations: the humanities help us look at design practices as composition, naming, narrating, and debating.
- They build an epistemological definition of design/conception: the aesthetic gaze and practice sever the connections between facts and theories, know-how and know-who, and therefore contributes to the dynamic flow of concepts and realizations.

---

[79] Ibid.

[80] In particular, Hatchuel et al. (2014).

7.6.3 *Pluri-Disciplinarity in Practice*

In each chapter, I have tried to show that my findings are consistent with design research trends. But I also wanted to produce, so to speak, an archaeology of the practices and concepts of design research by systematizing their origins in the bosom of two major trends of human sciences: the humanities and social sciences. To do so meant redistributing the attributes of a number of disciplines to be consistent with the analysis of design practices and objects. In this last section, I recall some of the disciplinary challenges that I encountered and how I tried to put in practice the indiscipline of design in the book.

Chapter 2 confronted two concepts of reflexivity: one that focuses on the phenomenology of our activity and that concentrates on dysfunctioning objects as an active ingredient of the reflexive stance, the other that focuses on stylistic properties of objects that supports reflexivity at the level of signs, texts, and media. The latter theories thus under-determine the phenomenological focus on the activity and its perception, while phenomenological considerations under-determine a strictly stylistic analysis of the designed object. The goal was to come up with a new definition of reflexivity which is grounded both in activity and media.

Chapter 3 showed how social sciences, whether sociology, cultural anthropology, and Human Computer Interactions, have augmented the theory of design by giving their insights into the users of artifacts. However, users/characters also appear in texts, discourses, graphic representations. In other words, they are part of narratives that exist in their own right to imagine the invention. I have not mentioned "personas" as none of the projects that I participated in used this design technique.[81] However, the notion of "figure of the user" seems to me to rearticulate the contributions of social sciences and the humanities to the concepts of users and personas in design because they consider the textual plane of creation as autonomous but dependent on the reader and consistent with her worldview.

In Chap. 4, I contrasted the sociology of science to the "poetics of science". The confrontation is not on the methods to study scientific activities but rather to consider how the latter are expansive and not only defensive or representative of a specific episteme. Linguistics with Jacobson and more broadly speaking literary and media studies helped me reconsider the contributions of poetic practices to invention in science. At the same time, poetic practices were also redefined since applied to science and thus part of an aesthetics of science.

In Chap. 5, the project as part of the epistemology of design and developed in disciplines ranging from management to ergonomics, was contrasted with composition as another axis of design activities. Both notions of project and composition are entangled but look at the same activity from different vintage points. To tackle the latter, I used the philosophy and history of art and applied arts to come up with definitions of composition as the organization of tensions. It is in this chapter that I refered to Peirce's abduction as a semiotic explanation for the properties of the

[81] Moggridge (2007).

composition but also to explain, thanks to Umberto Eco, the paradox of the apparent chaos of new compositions and their final coherence through the metaphoric process. This chapter is therefore indebted to semiotics and art to understand why we need to understand design from a spatial perspective.

Chapter 6 relied on translation and etymology to open a discussion on what it means to design "things" and not only "objects". This chapter tried to overcome the difference between the domain of discourses and the realm of objects. To get there I heavily relied on critical design and critical engineering research. Both posit that artifacts can foreground their values and therefore be critically examined. In other words, a semiotical analysis had to be articulated to a rhetorical and communicational analysis that brought the interactions but also the "thing" as part of the communicative action (Habermas[82]).

Chapter 7 is finally about stepping back and looking at the in-discipline of design. Not one discipline is concerned but potentially all of them when they engage into a conceptive path. A new perspective on the philosophy of sciences and the epistemology of conception within science had to be summoned and were mainly provided by Anne-Françoise Schmid.

7.6.4 *Last Word on In-Discipline: The Grace of the Heterogeneous*

I will again quote Phil Agre whose work on critical engineering was an inspiration to me. In his insightful text about Artificial Intelligence and the need for critical technical practice, Phil Agre, somehow dejectedly, acknowledged that to do both would not only be difficult but would almost require a split personality. "A critical technical practice will, at least for the foreseeable future, require a split identity -- one foot planted in the craft work of design and the other foot planted in the reflexive work of critique".[83] His personal life was unfortunately a testimony to that tragedy since he was bi-polar.

How can we tread this borderline that seems necessary not only to assess the technologies that we develop but to spur our creativity by questioning our presuppositions? Agre analyzed what would make this path both exciting and difficult. In particular, he acknowledged that one individual could not do it on her own but needed the support and environment where her position would be allowed and, in the best-case scenario, appreciated. Institutions need to develop spaces that not only question their scientific practices but expand them. In particular, the gap between sciences and critical studies can be bridged through a theory of design that lets us see what is conceptive in scientific practices because it includes the analytical techniques of the humanities.

---

[82] Habermas (1985).

[83] Agre (1997).

References

Agre, P. E. (1997). Toward a critical technical practice: Lessons learned in trying to reform AI. In G. Bowker, L. Gasser, L. Star & B. Turner, (Eds.), *Bridging the great divide: Social science, technical systems, and cooperative work*. Erlbaum.

Bourdieu, P. (1986). *Distinction: A social critique of the judgement of taste* (1st ed.). London: Routledge.

Bremner, C., & Rodgers, P. (2013). Design without discipline. *Design Issues, 29*(3), 4–13. doi:https://doi.org/10.1162/DESI_a_00217.

Bucciarelli, L. L. (1996). *Designing engineers*. Cambridge, MA: The MIT Press.

Caillois, R. (1992). *Les Jeux et les hommes : Le masque et le vertige*. Paris: Gallimard - Folio.

Cetina, K. K. (1999). *Epistemic cultures: How the sciences make knowledge*. Cambridge: Harvard University Press.

Chalmers, A. F. (1999). *What is this thing called Science?* (3rd ed.). Hackett Publishing Co., Indianapolis.

Coutellec, L. (2015). *La science au pluriel : Essai d'épistémologie pour des sciences impliquées*. Versailles: Quae éditions.

Cross, N. (2001). Designerly ways of knowing: Design discipline versus design science. *Design Issues, 17*(3), 49–55. doi:https://doi.org/10.1162/074793601750357196.

Cross, N. (2006). *Design as a discipline. Designerly Ways of Knowing*. London: Springer.

Csikszentmihalyi, M. (1991). *Flow: The psychology of optimal experience* (1st ed.). New York: Harper Perennial.

Dubreuil, L. (2007). Défauts de savoirs. *Labyrinthe, 27*, 13–26.

Dykes, T. H., Rodgers, P. A., & Smyth, M. (2009). Towards a new disciplinary framework for contemporary creative design practice. *CoDesign, 5*(2), 99–116. doi:https://doi.org/10.1080/15710880902910417.

Eco, U. (2004). *Mouse or rat: Translation as negotiation*. London: Phoenix.

Ewenstein, B., & Whyte, J. (2009). Knowledge practices in design: The role of visual representations as 'epistemic objects'. *Organization Studies, 30*(1), 07–30.

Findeli, A. (2006). Qu'appelle-t-on "théorie" en design ? Réflexions sur l'enseignement et la recherche en design. In B. Flamand (Ed.), *Le design : Essais sur des théories et des pratiques* (pp. 77–97). Paris: Editions du Regard.

Findeli A., Brouillet D., et alii (2008). Research Through Design and Transdisciplinarity: A Tentative Contribution to the Methodology of Design Research, in Aebersold R. et al., *« Focused » – Current Design Research Projects and Methods*. Genève: Swiss Design Network Symposium. 67–91.

Foucault, M. (1966). *Les mots et les choses; une archéologie des sciences humaines*. Paris: Gallimard.

Gaste, Y., & Gentes, A. (2013). Place and non-place: A model for the strategic design of place-centered services. *Bell Labs Technical Journal, 17*(4), 21–36. doi:https://doi.org/10.1002/bltj.21572.

Gentes, A. (2015a). Science fiction ? Scénarios narratifs et visuels dans les projets d'ingénierie des TIC. D. Dubuisson, S. Raux, & Collectif, *A perte de vue : Les nouveaux paradigmes du visuel*. Dijon: Les Presses du réel.

Gentes, A. (2015b, mai). Arts et sciences du design: la place des sciences humaines. *Sciences du design*, no 1, PUF, 96–109.

Gentes, A., & Jutant, C. (2012). Nouveaux médias aux musées. Le visiteur équipé. *Culture et Musées., 19*, 67–91.

Gentes, A., & Mollon, M. (2015). Critical design: A delicate balance between the thrill of the uncanny and the interrogation of the unknown. In D. Bihanic (Ed.), *Empowering users through design: Interdisciplinary studies and combined approaches for technological products and services* (pp. 79–101). New York: Springer.

Gentes, A., & Selker, T. (2013). Beyond Rhetoric to Poetics in IT Invention. In *Proceedings INTERACT 2013* (pp. 267–79).
Gentes, A., Valentin, F., & Brulé, E. (2015) Moodboards as the tool of the indiscipline of design. In *Proceedings IASDR, Brisbane* (pp. 755–771).
Guillory, J. (2010). Genesis of the media concept. *Critical Inquiry, 36*(2), 321–362. doi:https://doi.org/10.1086/648528.
Guy, B. (2015). Confrontation des démarches épistémologique et éthique du point de vue des sciences de l'ingénieur. In Y.-C. Lequin, P. Lamard, & Collectif (Eds.), *Eléments de démocratie technique*. Belfort: Université de Technologies de Belfort-Montbéliard.
Habermas, J. (1985). *The theory of communicative action, Volume 1: Reason and the rationalization of society* (trans: McCarthy, T.). Boston: Beacon Press.
Harfield, S. (2008). On the roots of undiscipline. In *Undisciplined! Design research society conference*. Sheffield Hallam University, Sheffield, UK, 16–19 July 2008.
Harris, R. (2000). *Rethinking writing*. Bloomington: Indiana University Press.
Hatchuel, A., Weil, B., & Collectif. (2014). *Les nouveaux régimes de la conception : Langages, théories, métiers*. Paris: Editions Hermann.
Higgins, D., & Higgins, H. (2001). Intermedia. *Leonardo, 34*(1), 49–54.
Latour, B. (2005). *La science en action : Introduction à la sociologie des sciences*. Paris: Editions La Découverte.
Legay, J.-M. (2004). L'interdisciplinarité vue et pratiquée par les chercheurs en Sciences de la vie. *Natures, Sciences, Sociétés, 12*, 63–74.
Legay, J.-M., & Schmid, A.-F. (2004). *Philosophie de l'interdisciplinarité : Correspondance (1999–204) sur la recherche scientifique, la modélisation et les objets complexes*. Paris: Editions Pétra.
Leigh Star, S. (2010). This is not a boundary object: Reflections on the origin of a concept. *Science, Technology & Human Values, 35*(5), 601–617. doi:https://doi.org/10.1177/0162243910377624.
Mackay, R. (Ed.). (2015). *Simulation, exercise, operations*. Falmouth: Urbanomic.
Mackay, W., & Fayard, A.-L. (1997). HCI, natural science and design: A framework for triangulation across disciplines. In *Proceedings of the 2nd conference on Designing interactive systems: Processes, practices, methods, and techniques* (pp. 223–234).
McKay, J., & Marshall, P. (2001). The dual imperatives of action research. *Information Technology & People, 14*(1), 46–59. https://doi.org/10.1108/09593840110384771.
McKay, J., & Marshall, P. (2007). Science, design, and design science: Seeking clarity to move design science forward in information systems. In *Proceedings of Australasian Conference on Information Systems* (pp. 604–614).
McLuhan, M. (1965). *Understanding media: The extensions of man*. New York: McGraw-Hill.
Méadel, C., Musiani, F., & Collectif. (2015). *Abécédaire des architectures distribuées*. Paris: Presses de l'Ecole des mines.
Meijers, A. W. M., et al. (2009). *Philosophy of technology and engineering sciences* (1st ed.). Amsterdam/Boston: North Holland: Elsevier.
Moggridge, B. (2007). *Designing interactions* (1st ed.). Cambridge, MA: The MIT Press.
Nieuwdorp, E. (2007). The Pervasive discourse: An analysis. *Computers in Entertainment, 5*(2), 13. doi:https://doi.org/10.1145/1279540.1279553.
Olohan, M. (Ed.). (2000). *Intercultural faultlines: Research models in translation studies: Textual and cognitive aspects* (Vol. 1). Manchester: St Jerome Publishing.
Perec, G. (2008). *Species of spaces and other pieces* (New ed.). London: Penguin Classics.
Rancière, J. (2006). Thinking between disciplines: An aesthetics of knowledge. *Parrhesia, 1*(1), 1–12.
Reiss, K., & Vermeer, H. J. (2013). *Towards a general theory of translational action: Skopos theory explained* (trans: Nord, C.). St. Jerome Publishing.
Rodgers, P., & Bremner, C. (2013). Exhausting discipline: Undisciplined and irresponsible design. *Architecture and Culture, 1*(1), 142–161. doi:https://doi.org/10.2752/175145213X13756908698720.

Schmid, A.-F. (1998a). *L'âge de l'épistémologie*. Paris: Editions Kimé.
Schmid, A.-F., (1998b). Une critique quasi-kantienne de l'épistémologie ou comment donner une positivité aux thèses de l'épistémologie », conférence, February 1998.
Schmid, A.-F. (2001). Pour une épistémologie de la conception. In Collectif & J. Perrin (Eds.), *Conception entre sciences et art: Regards multiples sur la conception* (pp. 79–87). Lausanne: Presses Polytechniques et Universitaires Romandes.
Schmid, A.-F., (2012a). Epistémologie générique et interdisciplines, in *Séminaire d'été, Maison des Sciences de l'Homme d'Aquitaine, Vers unescience de l'Europe ? L'interdisciplinarité dans le contexte du dialogue philosophique entre la France et la Russie*. Bordeaux, 3, 4 and 5 September 2012.
Schmid, A.-F., (2012b). SIG special interest group in design theory, January 2012, Paris, France.
Schmid, A.-F. (2015). On contemporary objects. In R. Mackay (Ed.), *Simulation, exercise, operations* (pp. 63–68). Falmouth: Urbanomic.
Schmid, A.-F., Mambrini-Doudet, M., & Hatchuel, A. (2011). Une nouvelle logique de l'interdisciplinarité. *Nouvelles perspectives en sciences sociales, 7*(1), 105–136.
Simatic, M., & Gentes, A. (2010). RFID-based distributed shared memory for pervasive games. In S. Ystad et al. (Eds.), *MobiCASE 2009, LNICST 35* (pp. 339–342).
Simatic, M., Astic, I., Aunis, C., Gentes, A., Guyot-Mbodji, A., Jutant, C., & Zaza, E. (2009). Plug: Secrets of the Museum: A pervasive game taking place in a museum. In *Proceedings of entertainment computing – ICEC 2009, Lecture notes in Computer Science* (pp. 67–74). Springer Verlag.
Simondon, G. (2001). *Du mode d'existence des objets techniques*. Paris: Aubier.
Souchier, E., Jeanneret, Y., & Le Marec, J. (2003). *Lire, écrire, récrire : Objets, signes et pratiques des médias informatisés*. Paris: Bibliothèque Publique d'Information.
Stein, Z. (2007). Modeling the demands of interdisciplinarity: Toward a framework for evaluating interdisciplinary endeavors. *Integral Review, 4*(1), 91–107.
Toury, G. (1995). *Descriptive Translation Studies and beyond*. Amersterdam/Philadelphia: John Benjamins Publishing Company.

Kolibri 360 GmbH
Ferdinand-Jühlke-Straße 7
99095 Erfurt, Deutschland
produktsicherheit@kolibri360.de